Learning Robotics with ROBOTIS DREAM Systems

By Chi N. Thai

CNT Robotics LLC, Lawrenceville

2018

Author: Chi N. Thai

roboteer@comcast.net

www.cntrobotics.com

Learning Robotics with ROBOTIS DREAM Systems (2018)

ISBN 978-0-9993918-1-5 (Paperback).

Library of Congress Control Number: 2018936047.

Published by CNT Robotics LLC, Lawrenceville, GA, U.S.A.

Printed by CreateSpace.

To my parents for giving me life.

To Christine and Emily for their enduring love and support.

To "Burrito" and all my grand nieces and nephews.

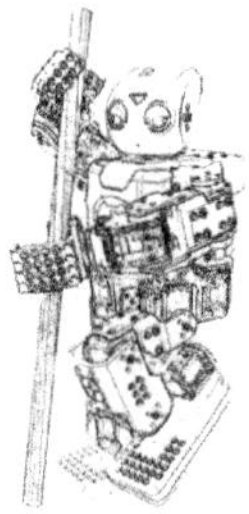

Preface

The DREAM II™ (School Set) programmable robotic kit was released by ROBOTIS® in Spring 2018 for the USA market with a cost around $220 US. It is recommended for users at age 8 or older. It comes with instructions to build 23 programmable robot examples and it can be interfaced with two free popular programming tools:

1) The first interface uses a ROBOTIS tool called TASK™ which can generate machine code that runs on the robot controller CM-150 allowing it to interact with its built-in NIR sensors and miniature speaker, along with a variety of external actuators and sensors. These TASK codes can be developed on MS Windows® platforms or on iOS® and Android® mobile devices, and they can be deployed via USB (wired) or **Bluetooth®**.

2) The second interface uses the Off-Line version of MIT's SCRATCH® 2 software to combine the power and multimedia services of a Windows PC with a direct control of the robot controller CM-150 via USB (wired) or Bluetooth and a helper application named R+SCRATCH, provided by ROBOTIS.

Who is this book for?

This book is for you if you are a young robotics enthusiast looking at achieving on your own a firm foundation in robotics design and programming, or if you are an adult investigating the possible use of the DREAM II School Set to help children learn about robotics programming and design. This book will show that this kit can be quite a versatile tool to introduce students from 8 to 12 years old to fundamental concepts in several areas: mechanical design, computer programming, robot control, inter-device communications and multimedia programming for richer story telling.

What do you need?

At the minimum, you need to purchase the ROBOTIS DREAM II School Set which is widely available via several web stores. Out of the box, this School Set is designed to be used within an MS Windows compatible environment for robot program development and download via a USB port using a ROBOTIS free tool called TASK V.2. It also includes a basic NIR-based remote control system (RC-100B).

The DREAM II School Set can also be used on mobile devices (phone or tablet), and the user would need to have the mobile App R+mTASK installed on these devices. Depending on the operating systems on your device, these Apps are available at (https://itunes.apple.com/us/developer/robotis-co-ltd/id948481761) or (https://play.google.com/store/apps/developer?id=ROBOTIS&hl=en). With this configuration, the user can create TASK programs that run on the Hardware Controller CM-150 and interact with various actuators and sensors. The user can configure the mobile device as a remote controller of the robot via a Bluetooth receiver named BT-410 (which is an additional purchase).

If the user wants to use this DREAM II kit with the SCRATCH 2 software, then the user is currently restricted to Windows PCs, as the needed helper software named R+SCRATCH is currently available for Windows OS only. Consequently, the user needs to use the Windows version of the SCRATCH 2 Off-Line Editor. This web site (http://www.robotis.us/software/) contains download links for all needed DREAM II software. The user can use the furnished micro USB cable to control the robot from the SCRATCH 2 IDE in a tethered mode. If the user wants the robot to have more freedom of movement, the user needs to additionally purchase the BT-410 receiver to put on the robot and also the BT-410 USB Dongle to be used on the PC side.

What is in this book?

This book consists of 6 chapters:

1) Chapter 1 presents an overview of the DREAM II system and its relationship with the SMART III system. The Sense-Think-Act paradigm used in developing the contents of this book is also described in this chapter.

2) Chapter 2 describes the hardware and software capabilities of the complete DREAM II system and shows how to get started with the School Set on Windows PCs as well as on Mobile Devices. This chapter also shows how to use the ROBOTIS MANAGER software tool using a basic wheeled robot design.

3) Chapter 3 is a substantial chapter providing a gradual but in-depth tutorial about applications of the R+TASK software tool using three robot designs - "Avoider/Follower", "TriCycle" and "Dowel Scanner". Topics included autonomous-behavior and remote-control algorithms, communications and audio programming techniques. PC and Mobile uses of the TASK tool are developed in this chapter.

4) Chapter 4 is also another substantial chapter mirroring the instructional approach and topics developed in Chapter 3 but now using the R+SCRATCH/SCRATCH 2 tool chain and its multimedia and event programming features.

5) Chapter 5 presents mechanical design concepts inherent in the mechanical components provided in Level 1 of the "complete" DREAM II system (i.e. non-programmable), with the goal of helping students understand the mechanical design concepts represented in the provided Level 1 example robots and be creative in their own robot designs by showcasing additional mechanical concepts and robot designs.

6) Chapter 6 provides a closer look at select programmable robots provided in the School Set (i.e. Levels 2 and 3) to explain their hardware/software features and to offer suggestions to expand some selected robots beyond their original designs or solutions.

This book also provides appropriate source codes and tutorial videos (via YouTube®) to illustrate the presented concepts, along with review questions to help students master learned materials. Please visit www.cntrobotics.com/dreambook for access options to the source codes and tutorial videos.

How should you use this book?

All readers should first read Chapters 1 and 2. Ideally, readers should practice the MANAGER related procedures as described in the text of Chapter 2 and its companion videos. This practice would require the use of either the USB cable or the USB BT-410 Dongle along with the BT-410 receiver.

Chapter 3 should be tackled next if readers are interested in using the TASK tool to program autonomous and remote-control behaviors into robots.

Chapter 4 is for users who are interested in using SCRATCH 2 and want to build their expertise in applying event and multimedia programming techniques to robotics.

Chapter 5 can be read independently from all other chapters however Chapter 6 requires mastery of the materials presented in Chapters 1 through 5.

And please, do let the author know about your experiences in using the materials in this book, this will help the author improve the next edition and the companion resources. Please visit the author web site www.cntrobotics.com for updates and other robotics materials. Please send your comments and suggestions to roboteer@comcast.net.

Trademark Attributions

Content in this book made references to the following trademarks:

ROBOTIS product names are trademarks or registered trademarks of ROBOTIS, Inc. Used with permission by ROBOTIS Co. Ltd.

Android®, YouTube®, and Chrome® are registered trademarks of Google Inc.

SCRATCH® is a registered trademark of the Scratch Team at the Massachusetts Institute of Technology.

Windows® is a registered trademark of Microsoft Inc.

iOS®, iPhone® and iPad® are registered trademarks of Apple Corporation.

Bluetooth® is a registered trademark of Bluetooth SIG, Inc.

Lawrenceville, GA, USA   Chi N. Thai

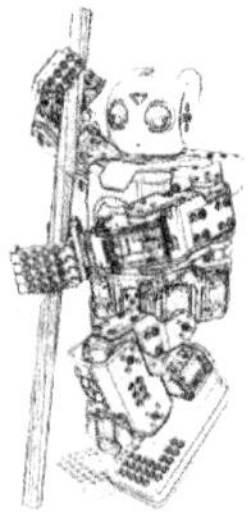

Contents

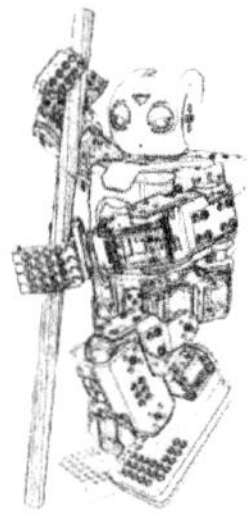

Chapter 1: Introduction

1.1 ROBOTIS DREAM II System Overview

ROBOTIS offers quite a large range of educational robotics kits to fit different user's ages and skill sets, please review this YouTube® video (https://youtu.be/Rxmgfose8wE) for a survey of the ROBOTIS "family" of robots.

For the USA market, ROBOTIS offers the PLAY and DREAM robotics systems for students in elementary and middle schools. The PLAY system uses the sturdier 7 mm DIA 1-component rivet system which is better suited to motor skill levels of younger children. The DREAM system uses a smaller 3.5 mm DIA 2-component rivet system which requires more visual/manual skills from the students (see Fig. 1.1). Please note that official ROBOTIS materials prefer to use the "distance between adjacent rivet holes" as the classification parameter, thus "3.5 mm DIA" corresponds to the "6 mm" hole distance, and "7 mm DIA" corresponds to the "12 mm" hole distance.

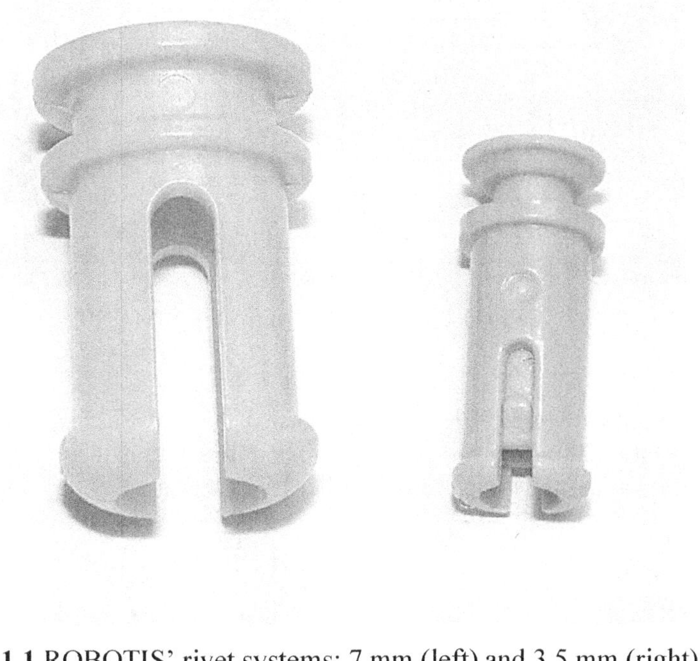

Fig. 1.1 ROBOTIS' rivet systems: 7 mm (left) and 3.5 mm (right).

The DREAM II system (http://www.robotis.us/dream/) offers an extensive range of sensors such as Touch, Color, Magnetic, Passive IR and Temperature sensors. It can operate different types of actuators such as continuous-turn and servo motors. It can be made to work with the R+m.PLAY700 App allowing the robot to access multimedia services on mobile devices (see Thai (2018)), or with MIT's SCRATCH 2 to achieve similar goals on PC platforms (this book).

The DREAM II system shares into a common ROBOTIS System Design Paradigm which considers a typical ROBOTIS robot to be a Computer Network with four major components:

1) A Hardware Controller, i.e. CM-150 (see Fig. 1.2), which is the "Main Brain" for the robot. It contains the user's instructions for the robot to execute the tasks programmed by the user.

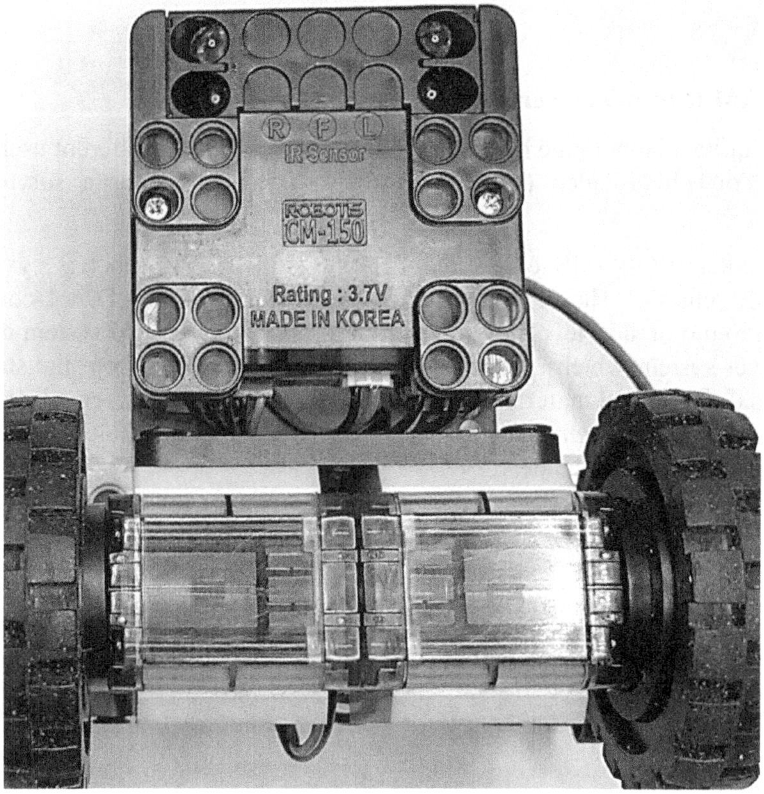

Fig. 1.2 Hardware Controller: CM-150.

2) The Sensors component (e.g. NIR or Touch sensors – see Fig. 1.3), which helps the Hardware Controller get information about its surroundings. The more sophisticated sensors have their own mini-brains to communicate with the Hardware Controller.

Fig. 1.3 Selected Sensors: NIR Sensor (left) and Touch Sensor (right).

3) The Actuators component (e.g. Servo Motors or LEDs – see Fig. 1.4), which allows the Hardware Controller to perform the appropriate robot actions upon the real world as programmed by the user. Similarly, advanced actuators such as servo motors have their own micro-controllers to work in concert with the Hardware Controller.

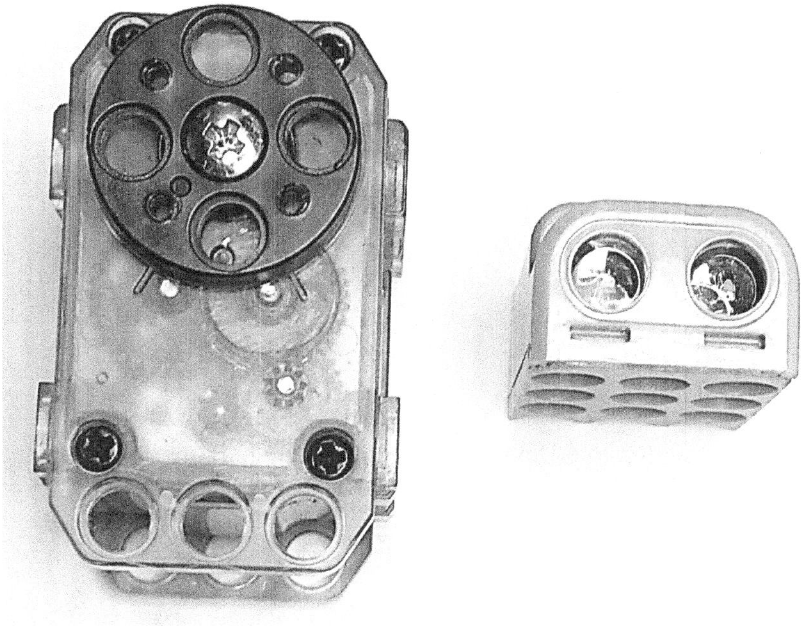

Fig. 1.4 Selected Actuators: Servo Motor (left) and LED Display (right)

4) A Remote Device, which can be an RC-100B module or a smartphone (see Fig. 1.5) or even a Windows PC. The Remote Device has a more flexible role depending on the specific device used and on its current role in this network of robotic components. For example, if the basic RC-100B is used with the DREAM system, it essentially functions as a NIR-based Sensor (attached to the user) informing the Hardware Controller about which "Up/Down/Left/Right" buttons had been pressed by the user. The basic RC-100B would require line-of-sight access between the physical RC-100B and the NIR receiver (IR-10) attached to the robot. When a smartphone (running the R+m.PLAY700 App) is used with the CM-150, it can act as a BlueTooth Sensor telling the CM-150 about which Touch Area on the phone screen had been pressed by the user, or it can act as an Actuator displaying appropriate graphics or videos as commanded by the CM-150 according to its programmed logic (see Thai (2018)). When a PC is used as a Remote Device, the issues become more complex. For example, if TASK's PC Virtual Controller (see Fig. 1.6) is used, then the PC acts as a glorified Sensor informing the Hardware Controller CM-150 about which keyboard buttons had been pushed by the user. However, if the SCRATCH 2/R+SCRATCH tool chain is instead used on the PC, then the PC takes on the role of a "true" Hardware Controller, as SCRATCH 2 codes run on the PC and not on the CM-150 controller which now has a diminished role, i.e. just letting SCRATCH 2 commands from the PC to pass through to Actuators and Sensors, via the CM-150's Firmware (see Fig. 1.8).

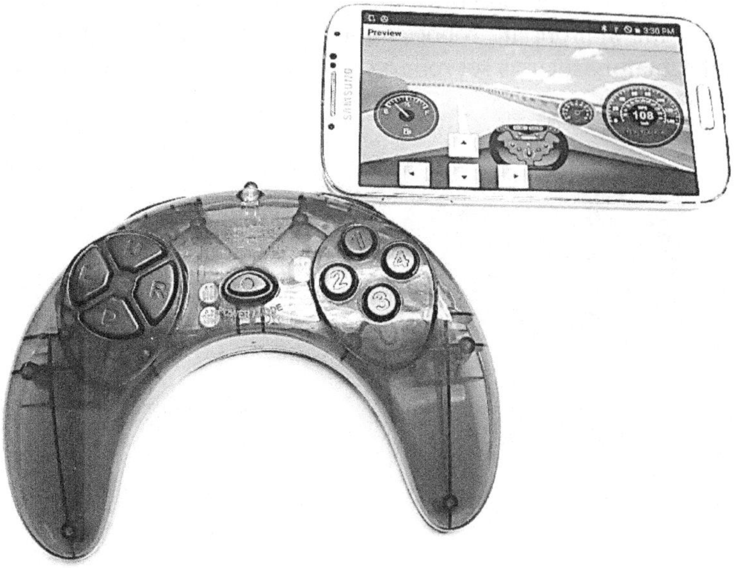

Fig. 1.5 Selected Remote Devices: RC-100B (Left) and Smartphone (right).

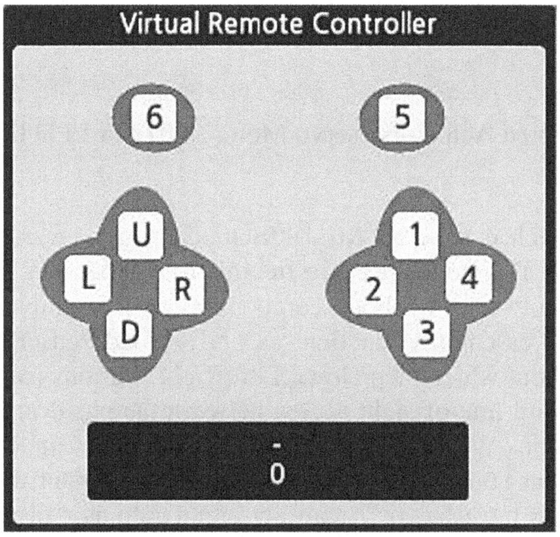

Fig. 1.6 TASK's PC-based Virtual Controller Interface.

To achieve hierarchical control in this adaptive network of robotic components, **ROBOTIS software uses two types of communication packets: Instruction and Status**. Instruction packets can only be issued by the Top-Level Controller which can be either the Hardware Controller (CM-150) or the Remote Device, and there can be only ONE Top-Level Controller at any one time (if ones use the proprietary firmware provided by ROBOTIS for its controllers). Once a network component receives an Instruction packet, it will execute the prescribed instruction and will optionally send back Status packets to the originator of that Instruction packet.

Fig. 1.7 shows a configuration of network components and communication packets flow for the situation where the Top-Level Controller is the Hardware Controller (CM-150) using the RoboPlus software suites (i.e. MANAGER, TASK, MOTION), while the Remote Device could be any of the following devices: physical RC-100, virtual RC-100 on the PC, smartphone or tablet running the R+m.PLAY700 App. In this configuration, please note that only the Hardware Controller can issue Instruction packets and that Actuators and Sensors can never issue Instruction packets as they are at the "bottom" of the control hierarchy.

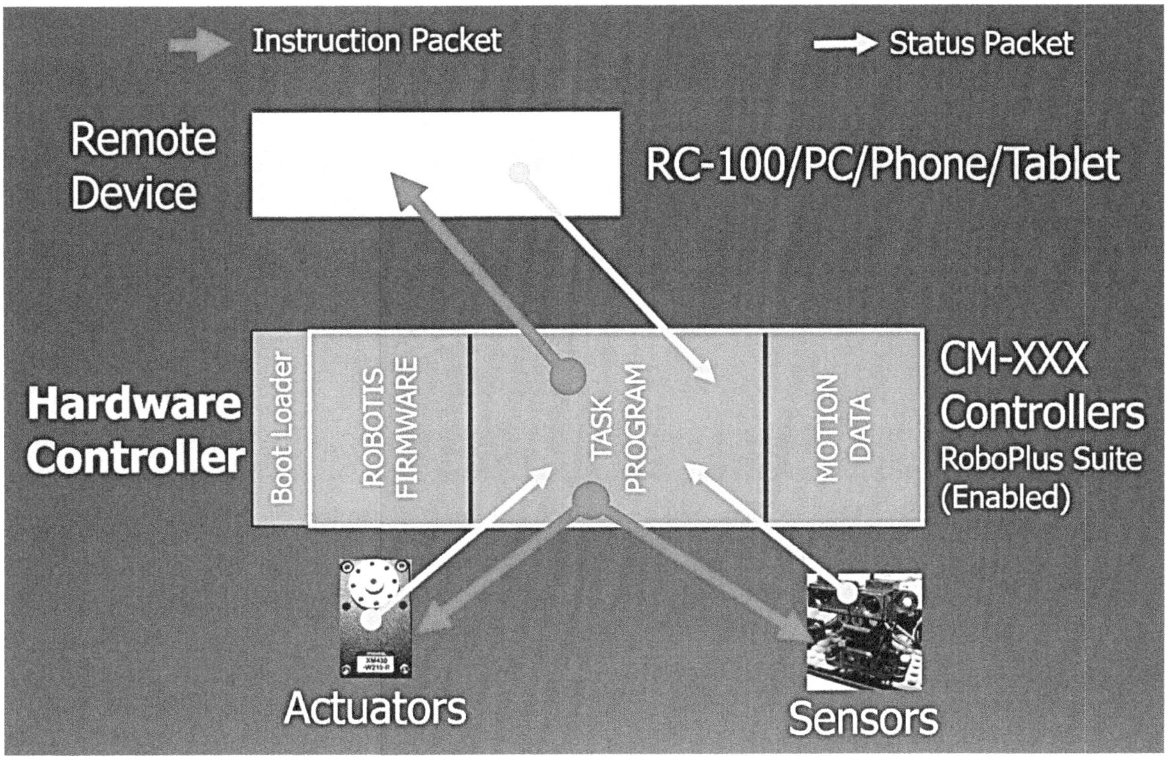

Fig. 1.7 Hardware Controller as Top-Level Controller.

Fig. 1.8 shows a second network configuration where the Top-Level Controller is the Remote Device which can be a PC using the SCRATCH 2/R+SCRATCH tool chain or a mobile device using a more advanced ROBOTIS software tool called OLLOBOT SDK (which is beyond the scope of this book and interested readers are referred to Thai (2017) for more details). In this configuration, please note that all TASK and MOTION functionalities would now be disabled on the CM-150 (mainly because ROBOTIS controllers' firmware are written this way). Please also note that all communication packets are now routed through the R+SCRATCH application on the PC and the "Firmware" component of the controller before reaching the Actuators and Sensors, meaning that communication delays would likely occur for this configuration due to the extra layer of communication software (especially when serial communication through R+SCRATCH is restricted to a rate of 57.6 Kbps).

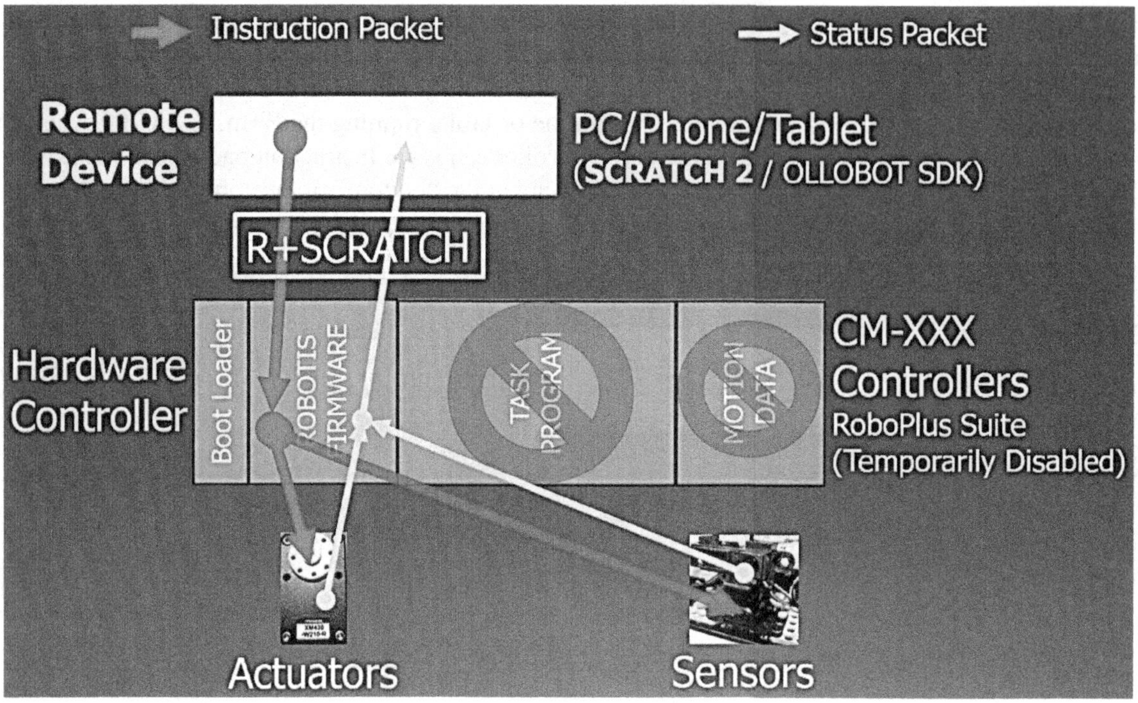

Fig. 1.8 Remote Device as Top-Level Controller.

There is a third configuration for a more advanced and flexible use of ROBOTIS robotic systems with Embedded C, which unfortunately is outside the scope of this book but interested readers can refer to this web link for more details (https://www.cntrobotics.com/robotis-evolving-dynamixel-concept). Advanced readers can learn more about Instruction and Status packets at (http://support.robotis.com/en/product/actuator/dynamixel_pro/communication/instruction_status_packet.htm). In this book, we'll be using the software tools TASK and SCRATCH 2 wherein the task of handling Instruction and Status packets is performed by the Controller's Firmware **"behind the scene"**, thus fortunately readers won't have to deal with them directly.

Please note that there is a third type of packet, called "RemoCon", used in Remote Control schemes whereas the user will have the opportunity to have a more hands-on practice in Section 3.6.

1.1.1 DREAM II - Levels 1-5 Edition

Since Spring 2018, the complete DREAM II System is available for the U.S.A. market and it includes 5 levels (http://www.robotis.us/dream/) which work in a cumulative scheme: i.e. users would have to purchase/start with Level 1, then go to Levels 2 through 5 consecutively:

1) Besides the various frame parts, the major components of Level 1 include 1 Low-Speed Geared Motor, 1 Micro-USB cable and 1 LiPo battery unit. Level 1 showcases 11 motorized but non-programmable robots.

2) Level 2 adds 1 CM-150 Controller, 2 Geared Motors, 2 Rubber Tires and more frame parts. Level 2 showcases 11 programmable robots, some are wheel-based and some are linkage-based (i.e. "walking" robots) and all as autonomous behavior applications.

3) Level 3 adds 1 Servo Motor, 1 Remote Controller RC-100B and its NIR Receiver IR-10, and more frame parts. Level 3 showcases another 11 programmable robots using the Servo Motor and illustrating Remote Control techniques. Thus, the Classroom Set offers 33 example robot designs.

4) Level 4 adds 2 more Servo Motors, 2 External NIR Sensors, 2 Touch Sensors, 1 LED Module, and of course more frame parts. Level 4 showcases 12 additional robots with more sophisticated mechanical designs and behaviors, both autonomous and remote controlled.

5) Level 5 adds a set of Master and Slave BlueTooth modules to be used with the RC-100B, 1 more LiPo battery unit, and even more frame parts, including a Track system. Level 5 showcases 12 final wheel-based and track-based robots with autonomous and remote-controlled designs. Some applications use the more sophisticated Callback function which is a type of hardware-timer interrupt function.

1.1.2 DREAM II - School Set

The School Set is customized for the USA market and roughly corresponds to a bundle of the major components of Levels 1, 2 and 3 combined. Thus, both autonomous and remote-control applications can be achieved with the School Set. This book is specifically written for the School Set.

Fig. 1.9 DREAM II School Set (Courtesy of ROBOTIS).

1.1.3 From DREAM II to SMART III

As the SMART III system (http://www.robotis.com/shop/list.php?ca_id=103020) also uses the smaller 3.5 mm rivet system, ROBOTIS offers the SMART III Level 0.5 as an upgrade path to the CM-200 controller (http://www.robotis.com/shop/item.php?it_id=901-0054-102) for users who already own the DREAM I or II "complete" kits. The CM-200 controller (http://support.robotis.com/en/product/controller/cm-200_main.htm) can accommodate up to 8 servo motors and can fully utilize ROBOTIS MOTION V.2 tool (http://support.robotis.com/en/software/roboplus2/r+motion2/rplus_motion2.htm) (which is not supported on the DREAM system). Currently, Spring 2018, the SMART III system is only available in South Korea.

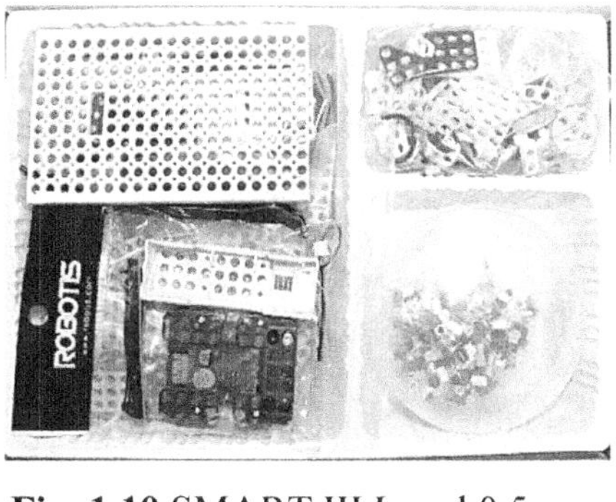

Fig. 1.10 SMART III Level 0.5.

1.2 Sense-Think-Act Paradigm

Usually the younger robotics students were surprised when I started my robotics short courses by mentioning that they had been using the Sense-Think-Act paradigm in everyday activities by themselves. I used Fig. 1.11 to explain how humans use our Senses to let us know about the external World (i.e. Perception) and apply those Sensations into our Cognitive process (i.e. Thinking) to devise proper Actions/Reactions to the situation at hand. These Actions/Reactions in turn would change some aspects of the external World resulting in new Sensations triggering the next World-Human-Interactions cycle and so forth.

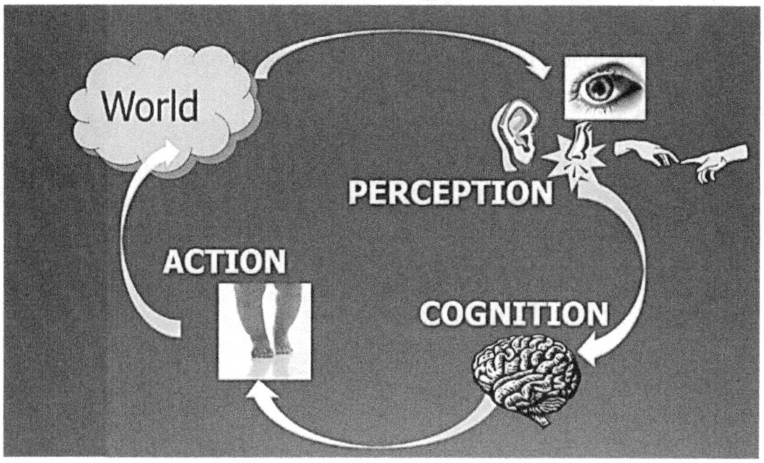

Fig. 1.11 World-Human Interactions Cycle.

Thus, in a way, doing Robotics is just trying to reproduce this interactions cycle onto a robot as shown in Fig. 1.12. Mechanical or electronic Sensors would convert their interactions with the World into Input Data to be sent to the Robot/Computer which was previously Programmed by the user to deal with Events deemed important to the operation of the Robot. Using this programmed Logic, the Robot Outputs appropriate commands to its Actuators which then could change the Sensors' perspective onto the World and therefore generate new Input Data for the Robot/Computer to process anew, through another interactions cycle.

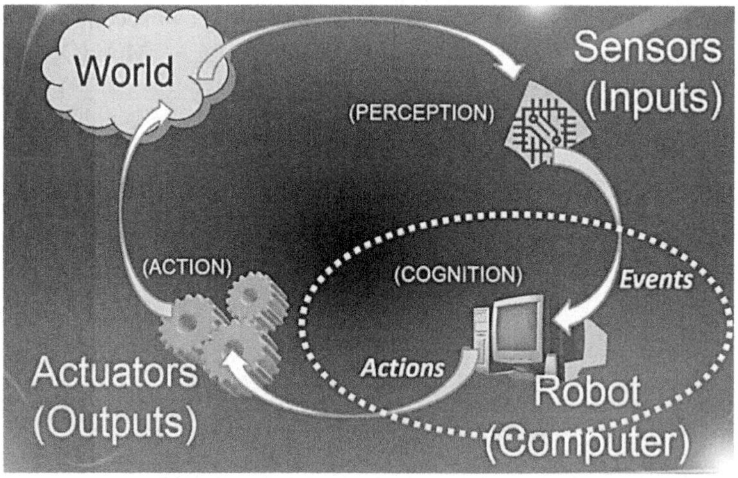

Fig. 1.12 World-Robot Interactions Cycle.

Let's go through an example to see how this Sense-Think-Act paradigm can help us solve the following challenge as described in Fig. 1.13, where a CM-150 based Probe Car robot (*which is in Level 4*) would be able to follow an irregular black track in the shape of the number 8. Furthermore, if the Probe Car encounters a frontal obstacle (e.g. a paper cup) during its travel, it would need to stop in its track, pick up the paper cup, move it out of the way and finally return to its original direction of travel (see Video 1.1).

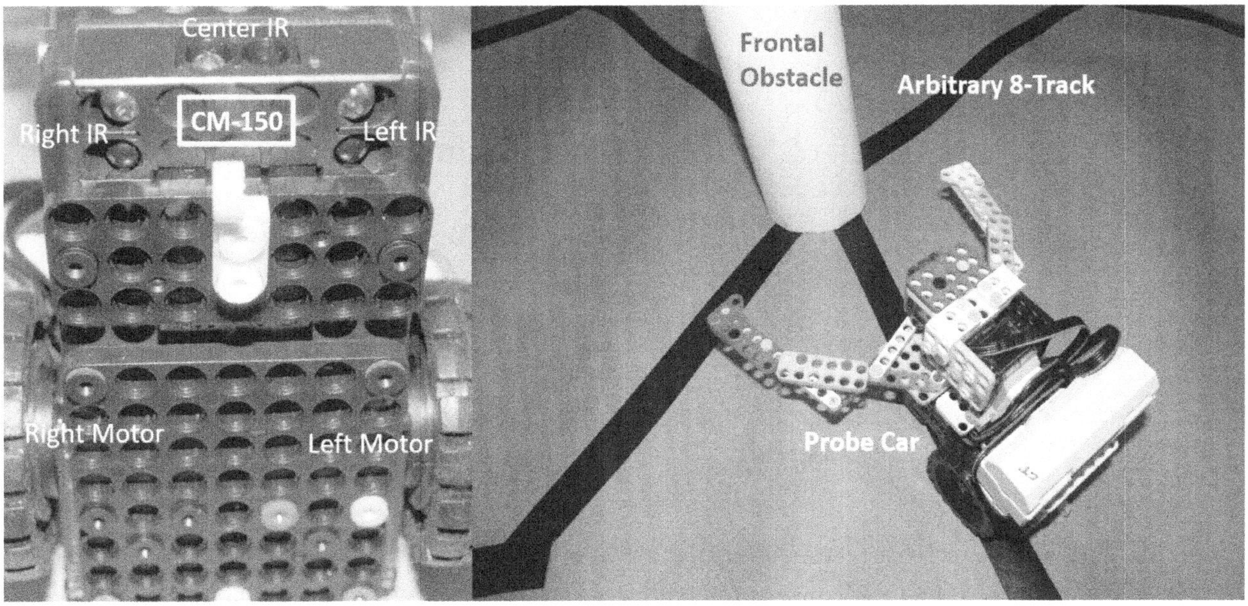

Fig. 1.13 CM-150 Probe Car following an 8-track and avoiding a frontal obstacle.

Fig. 1.13 shows that the CM-150 has three IR Sensors whereas the Center IR could be used to detect a frontal obstacle and the Right and Left IR sensors could be used to detect the curvy black track. The CM-150 also has a Motor on each Left and Right side, so these motors should be able to be programmed to provide all Probe Car maneuvers that would be needed to perform this challenge successfully (i.e. Forward, Turn Left and Turn Right). This System Design solution seems feasible enough, but to progress further towards the actual coding of the CM-150, we need to zoom in around the Robot/Computer area in Fig. 1.12 (i.e. the dashed yellow boundary in the lower right corner). The Sense-Think-Act paradigm still applies at this level, but some robotics practitioners would start using different labels such as "Events", "Actions" and "Reactive Control" as they are more useful in the next steps that we need to take in creating the actual computer code to control this Probe Car. Interested readers should read Matarić (2007) for a more complete description of other robotics control approaches, or Ben-Ari and Mondada (2018) for more details about "Reactive Control" and "Finite State Machines".

With the "Reactive Control" approach, ones need to match up a possible Condition/Event with an appropriate Action for the robot to do (see Fig. 1.14).

Reactive Control Approach

Given Condition (Event) >> Appropriate Robot Action

Condition/Event	Action
Car on top of Track	Car Goes Forward
Car left of Track	Car Turns Right
Car right of Track	Car Turns Left
Paper Cup detected	Pick Up, Dispose of Paper Cup & Rotate Left until Original Direction

1 and Only 1 Condition (Event) can happens at any 1 time.
Multiple Conditions (Events) can happen at any 1 time.

Fig. 1.14 Reactive Control Approach used for the Probe Car challenge.

For example, if the Probe Car happens to be lined up on top of the track already, then it should just go forward. There are 3 other Event/Action pairs listed in Fig. 1.14, and are these four pairs the only ones to consider or needed for this robot challenge? My advice for beginning roboteers is just to start with a reasonable set of Event/Action pairs and then improve on them as needed in later refinement steps.

But how would a beginner in robotics comes up with such a list as shown in Fig. 1.14 in the first place? Here, in keeping with the introductory scope of this book, the author offers a simple "thought experiment" or "visualization exercise" approach:

1) First, we would imagine to be "one" with the robot, so our Left and Right Eyes would match with the Left and Right IR Sensors, i.e. we keep our nose to the ground so that we can see the black track (to be exact, only a partial view of it). However, humans do not have a third eye right on top of our head to correspond to the Center IR, so we would really have to use our imagination and use our "mind's eye" as the third eye there!

2) So now that we are the robot, how would we go about performing this 8-track challenge? Here, although not mentioned in the challenge write-up, we need to decide on where is the starting position of the robot with respect to the 8-track? At the start of the challenge, the robot could very well be positioned on a brown paper area inside or outside the track, or it could straddle the track itself. As the challenge's rules did not specify the starting event, we are going to choose the "easiest" option of having the Probe Car start its challenge already lined up with the track. Then, the reader can see why the first Event/Action pair came about on the list in Fig. 1.14. Next, as the 8-track would curve here and there, the robot may find itself left or right of the track (second and third Event/Action pair respectively). So, with the first three Event/Action pairs, we should feel confident, at least logic wise, that the Probe Car would be able to follow the 8-track using computer codes derived from Fig. 1.14.

3) Similarly, we can come up with the 4th Event/Action pair to take care of the situation when an obstacle comes within detection range of the Center IR. Here the alert reader would notice that the first three Actions were Simple Actions triggered when certain conditions occurred. But the 4th one had several Actions in sequence and its last Action was a much more complicated "Action-until-Event", i.e. Act now until some Event happens later (e.g. Rotate Left until Probe Car returns to its original direction). In other words, there is a sort of a nested Event/Action inside an "Action-until-Event".

Before we go on to the next step, we need to have a closer look at the four Event/Action pairs (or Conditions if you will) listed in Fig. 1.14. Logically, the reader can see that the first three conditions are mutually exclusive (i.e. one and only one of them should happen at any one time in "real life"), but that the 4th condition may happen in conjunction with any of the first three conditions. Or using a different viewpoint, it is only when the 4th condition is false (i.e. no obstacle in front of the Probe Car) that we would care about how to handle the first three conditions. **So, the "take-home" lesson is that not all conditions "discovered" during this step have the same logical importance or ranking.** Please keep this hierarchy of conditions in mind when we get to the later steps of creating the actual TASK code for this challenge in Chapter 3.

The next step is to build upon the results of the previous step by mirroring its Event/Action pair into a matching Sensor/Actuator pair as shown in Fig. 1.15, as an Input/Output table. **The goal is to specify the kind of Sensors/Actuators to be used and to prescribe how they are to be used.**

Input-Output Table

Given Input Sensor(s) >> Activate Appro. Actuator(s)

Input Sensors	Output Actuators
Right IR & Left IR see "Black"	Right Motor CW Left Motor CCW
Right IR sees "Black" & Left IR sees "Brown"	Right Motor Stops Left Motor CCW
Right IR sees "Brown" & Left IR sees "Black"	Right Motor CW Left Motor Stops
Center IR sees "Obstacle"	Right Motor CW & Left Motor CW until Left IR sees "Black"?

Fig. 1.15 Input/Output Table mirroring list of Event/Action pairs as shown in Fig. 1.14.

For example, let's look at the second Sensor/Actuator pair in Fig. 1.15 which corresponds to the second Event/Action pair in Fig. 1.14 which specifies that if the Probe Car is found on the left of the black track then it should turn right so that it can get back onto the track. Thus, for the Probe Car to be on the left of the track, its Left IR should be seeing "Brown" and its Right IR should be seeing "Black" (see 2nd row of the Input Sensors column in Fig. 1.15). The corresponding Action for the Probe Car is then to turn right, i.e. the Right motor should stop and the Left motor should keep on going "Forward", i.e. keep on turning CCW (see 2nd row of the Output Actuators column in Fig. 1.15).

As an exercise to the reader, let's have a closer look at the specifications for the Output Actuators for the 4th Event when a frontal obstacle comes within the detection range of the Center IR Sensor. This 4th Event would trigger a sequence of complex actions as described in Fig. 1.14, but for lack of space in Fig. 1.15, we will just consider the very last Action "Rotate Left until Back on Track": i.e. the Probe Car is supposed to spin Left until it straddles the black track again and in the original travel direction, so the specifications were to run the Right motor "Forward" (i.e. CW) and to run the Left motor going "Backward" (i.e. CW also) so that the Probe Car would rotate Left, until the Left IR Sensor sees "Black". Or would the reader think that it should really be "until the Right IR Sensor sees Black" instead, to make sure that the Probe Car is completely straddling the black track? The answer will be given in Chapter 3.

Please note that the previous procedures do not involve the use of any computer nor learning about how to use a specific software tool, and the author highly recommends that beginner roboteers to always start with these procedures instead of jumping directly into coding or try to adapt someone else codes, as these procedures would help the roboteer refine in his/her mind the tasks to be done by the robot, before committing them into hard codes, thus saving the programmer lots of time in the long run.

Another benefit of using the Event/Action and Input/Output tables is that they will help with the actual writing of the TASK code (as shown later in Chapter 3) as the ROBOTIS syntax used would support practically a one-to-one correspondence of all Event/Action pairs A to Z from these tables into similar "physical" locations within the final program, except for two aspects that had not been discussed yet (see Fig. 1.16):

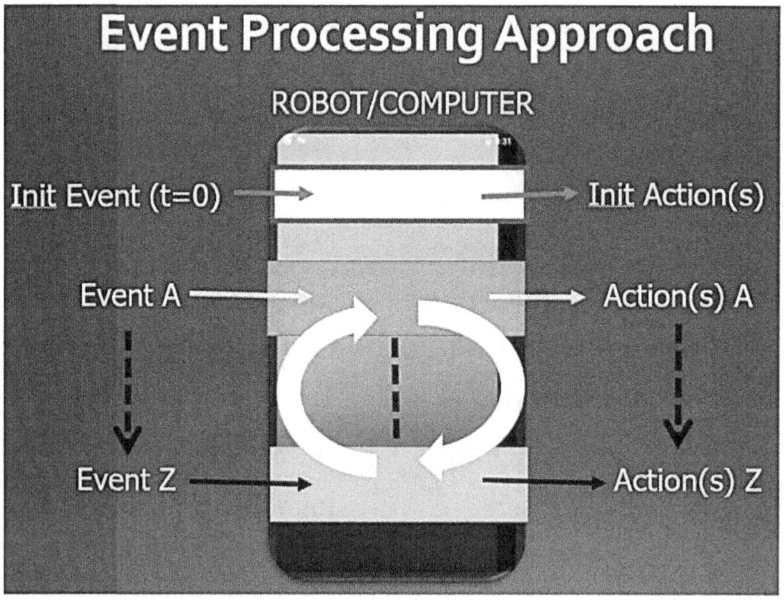

Fig. 1.16 Event/Action pairs matching locations in TASK code.

1) A main code looping structure that allows the controller to go through the list of possible Events A to Z that may occur during run time, as it listens to what the Sensors tell it about the current state of the World. This is the overall loop (i.e. Interactions Cycle) shown in Figs. 1.11 and 1.12.

2) The Init Event which occurs once at time=0, i.e. when power is applied to the robot or when the TASK program starts running. The Init Event would usually result in a known state for the robot which could be some audio output to let the user knows that it is ready for "action", or it could be a display of a

menu of tasks to let the user choose which task for the robot to perform when using R+m.PLAY700 App (see Thai (2018)).

In the next Chapter 2, we'll get into the details of setting up the hardware and software for the DREAM II School Set and show the readers how to use various software tools on Windows PCs and mobile devices.

1.3 References

Ben-Ari M, Mondada F (2018) Elements of Robotics. Springer Publishers International, Cham, Switzerland.

Matarić MJ (2007) The robotics primer. The MIT Press, Cambridge.

Thai CN (2017) Exploring Robotics with ROBOTIS Systems (2nd Edition). Springer Publishers International, New York.

Thai CN (2018) Advanced Application Programming with ROBOTIS DREAM System. Amazon Kindle Direct Publishing.

1.4 Review Questions for Chapter 1

1. What are the sizes of the rivet systems used on the DREAM and PLAY robotics systems?
2. The DREAM II Level 1 robotics kit is programmable. T F
3. The DREAM II Classroom Set includes servo motors T F
4. How many NIR Sensors are built-in on the CM-150 controller?
5. What is the name of the mobile software App that can be used to program DREAM II robots?
6. Does the DREAM system work with the R+m.PLAY700 software tool?
7. Name the hardware controller used in the SMART system.
8. Name the hardware controller used in the DREAM system.
9. List examples of Sensors for ROBOTIS systems.
10. List examples of Actuators for ROBOTIS systems.
11. List devices that can be used as Remote Controllers for the DREAM II system.
12. List types of communication packets used in ROBOTIS systems.
13. Describe the "Sense-Think-Act" paradigm.
14. Describe the Reactive Control approach.
15. List the hardware needed to use BlueTooth on a Windows PC platform.

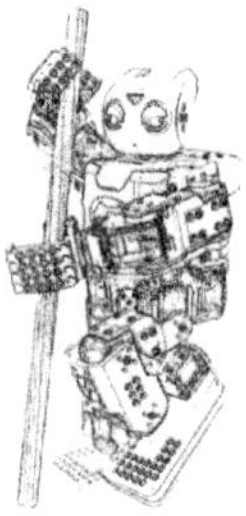

Chapter 2: Getting Started with DREAM II

Out of its shipping box, the DREAM II School Set is fully functional on a Windows PC platform equipped with a USB port using the provided micro USB cable. This means that the robot will be tethered to the PC via this 3-foot cable during code development, which should be satisfactory for most applications. Once robot programming is finalized, the user can detach the USB cable and use the RC-100B/IR-10 combination for NIR remote control (line of sight required and with a practical range of about 3-6 feet).

If a farther control range is needed (up to 30 feet), the user needs to invest in a set of Master-Slave BlueTooth modules which do not require "line-of-sight" and also have reliable "one-to-one" connections within a multiple-robot environment like in a robot competition (http://www.robotis.us/bt-410-set/). The "Master" BT-410 module needs to be attached to the RC-100B remote controller, while the unmarked "Slave" BT-410 module needs to be used on the robot. The "Master" and "Slave" modules are automatically and uniquely paired for the first time when powered within a few inches of each other.

Furthermore, if the user wants to develop and deploy robot codes wirelessly from the PC, then the user needs to also invest in the USB BT-410 Dongle (http://www.robotis.us/bt-410-dongle/), to be connected to an available USB port on the PC. The BT-410 Dongle is a BlueTooth bridge so it can pair to "any" Slave BT-410 module when they are in proximity of each other (1-2 inches). Once paired, they stay in "one-to-one" connection with each other up to 30 feet.

The software packages that the DREAM II kit can use are listed below:

1) R+DESIGN is a 3D CAD tool for users to aid in their assembly of existing ROBOTIS example kits, as well for experienced users to document their own designs using existing ROBOTIS components (http://support.robotis.com/en/software/roboplus2/r+design/r+design.htm).

2) R+MANAGER is used to check on the overall operation of the CM-150 controller and its built-in NIR and Sound Sensors, the external Geared Motors and to some extent the external Servo Motors (more details in Section 2.1). MANAGER can be used to update or recover the firmware on the CM-150 and its peripherals. The MANAGER tool only works on PCs.

3) R+TASK is the software tool used to develop the robot codes and to deploy them onto the CM-150 (see Chapter 3 for a more detailed treatment). A mobile version is available for iOS and Android devices and it is called R+m.TASK.

4) R+SCRATCH is a helper software serving as an interface between MIT's SCRATCH 2 software and the firmware residing on the CM-150 (see Chapter 4 for a more detailed treatment). R+SCRATCH only works on PCs.

This web site (http://www.robotis.us/software/) contains download links for all needed DREAM II software tools.

Next, let's build a Basic Avoider robot as the demonstration platform (see Fig. 2.1) for various software demonstrations in the next sections. Please watch Video 2.1 to see how to use R+DESIGN on the PC to put it together (only up to Step 201 of the DREAM Avoider design).

Fig. 2.1 Basic Avoider demonstration platform.

2.1 Using Basic Avoider on Windows PCs

Assuming the reader had installed successfully the following software packages on the PC:

1) MANAGER (http://www.robotis.us/roboplus2/).

2) TASK (http://www.robotis.us/roboplus2/).

3) R+SCRATCH (http://www.robotis.us/roboplus2/).

4) MIT SCRATCH 2 (https://scratch.mit.edu/download).

The MANAGER tool is used when the user needs to check if the CM-150 Controller and the actuators or sensors attached to it are working properly. Fig. 2.2 shows the main menu for the MANAGER tool when it connects to a CM-150. The column labeled "Control Table" contains various functions that can be tested directly via MANAGER: for example, Sound Detected Count, Ports 1 & 2 Motor Speeds, etc... These functions (and more) can also be programmed via the TASK tool (to be shown later in Chapter 3). Sometimes, the CM-150 may malfunction in some way, and usually the best solution is to "recover or update its firmware" which can only be done via this MANAGER tool. Please see Video 2.2 for more details.

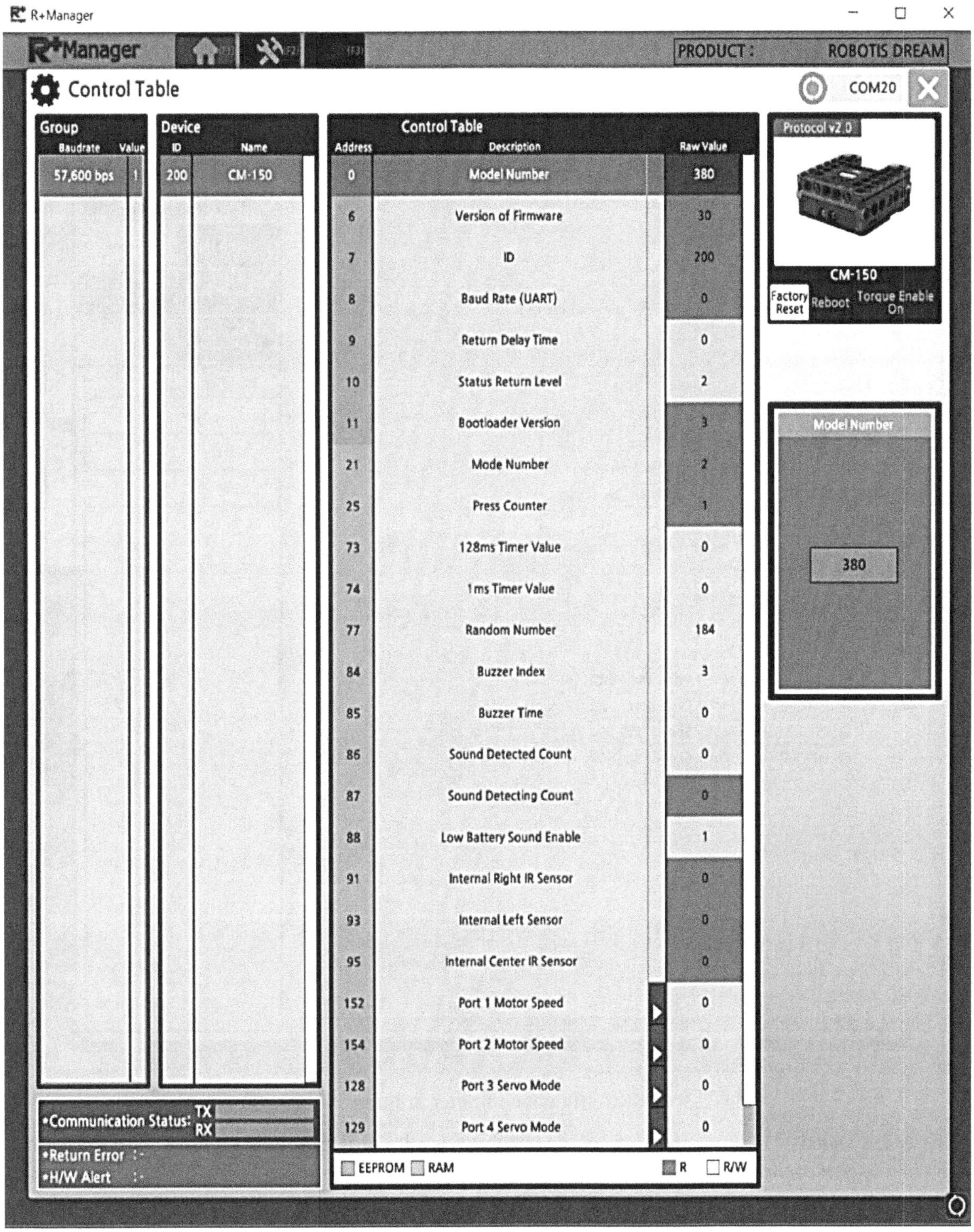

Fig. 2.2 MANAGER Tool on CM-150.

The TASK tool is THE "programming" tool, where the results from Input/Output tables, created from steps described in Chapter 1, are converted into TASK program statements. The TASK tool may be described as a context-sensitive line editor (see Fig. 2.3).

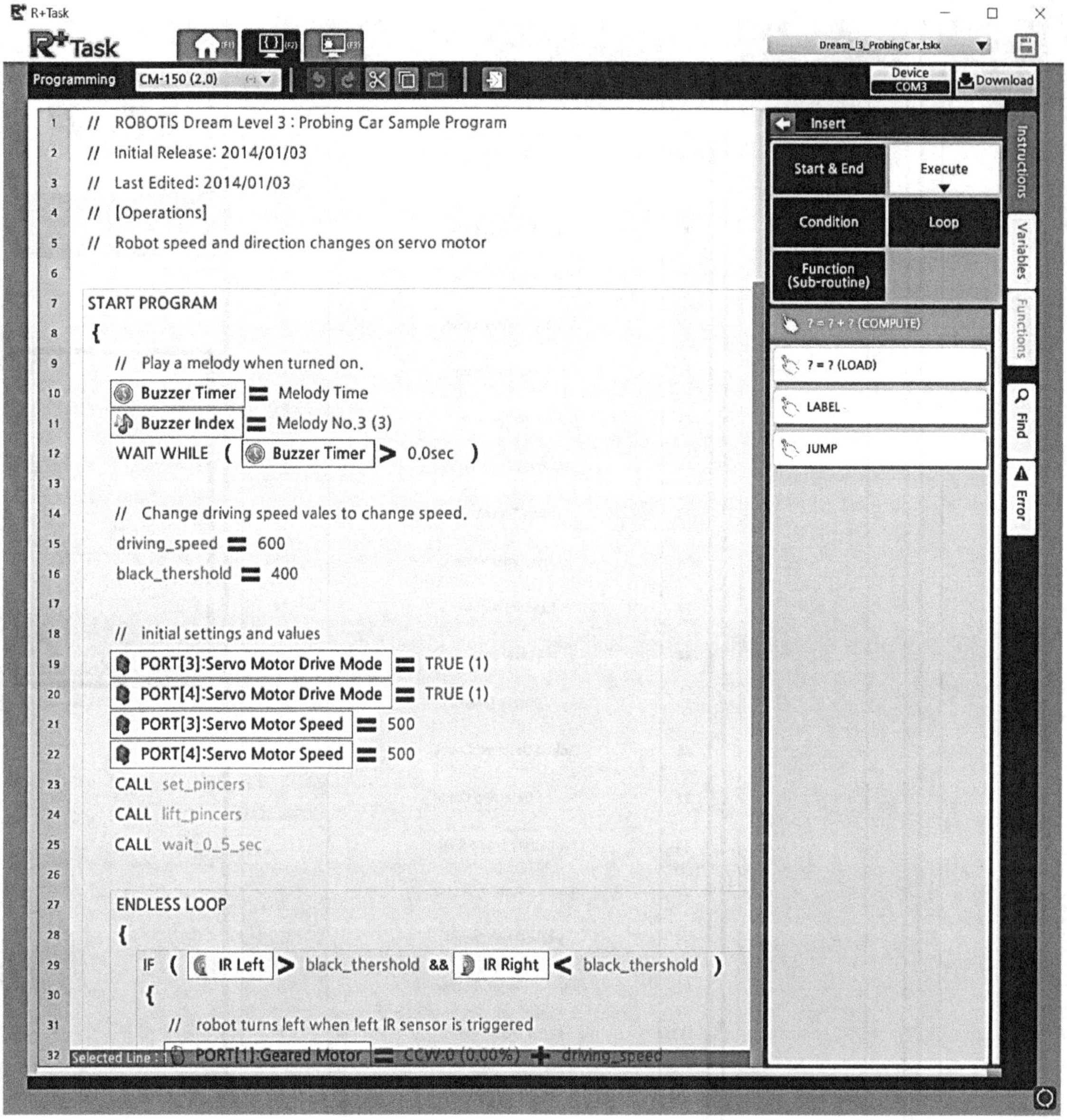

Fig. 2.3 TASK tool on a computer program for CM-150.

Please see Video 2.3 for a quick tour of its capabilities. Its proficient use requires lots of training and practice on the part of the user which will be developed in Chapter 3.

R+SCRATCH (Fig. 2.4) is an important background application that interfaces with the SCRATCH 2 software tool, to allow SCRATCH 2 commands to be sent from a PC communication port to reach into the actuators and sensors on the CM-150. Please note that the Firmware on the CM-50 needs to be V.31 or above for R+SCRATCH/SCRATCH2 to work properly with the DREAM II System. Please watch Video 2.4 for a demonstration of its typical use.

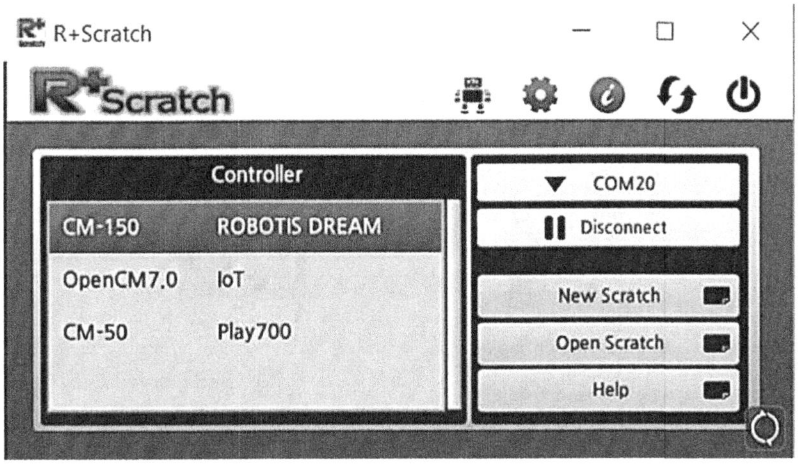

Fig. 2.4 R+SCRATCH background application for PC.

SCRATCH 2 is a well-known software tool with millions of users world-wide (https://wiki.scratch.mit.edu/wiki/Scratch_Statistics#Scratchers_Worldwide). The SCRATCH 2 tool may be described as a block-programming tool (see Fig. 2.5). Please see Video 2.5 for a quick tour of its event-programming capabilities which are most applicable to robotics. Its proficient use also requires lots of training and practice on the part of the user which will be developed in Chapter 4.

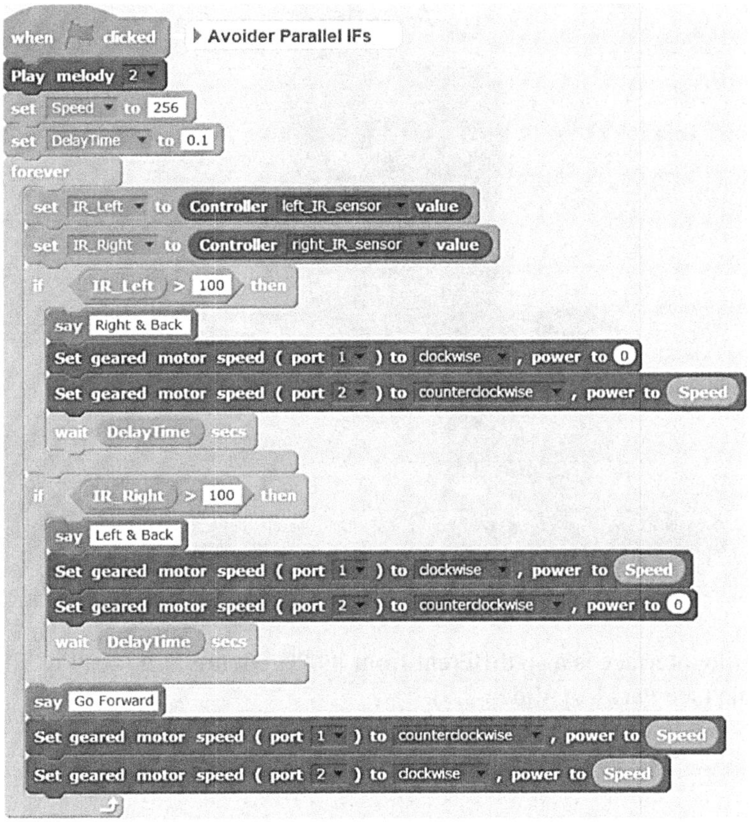

Fig. 2.5 SCRATCH 2 for Basic Avoider project.

2.2 Using Basic Avoider on Mobile Devices

For iOS and Android devices, currently there are 4 ROBOTIS Apps (available in English) that pertain to the DREAM and SMART systems: R+m.DESIGN, R+m.TASK, R+m.PLAY700 and R+m.MOTION (see Fig. 2.6).

Fig. 2.6 Four ROBOTIS Mobile Apps for DREAM and SMART systems.

R+m.MOTION2 is applicable to the SMART, BIOLOID and MINI systems only.

R+m.PLAY700 was originally created to work with the PLAY700 kit only, but the author had found that it can work with the DREAM system as well (see Thai (2018)).

R+m.DESIGN's mobile interface is different from the PC version as previously shown in Video 2.1 (see Fig. 2.7), however its functionality is similar (see Video 2.6).

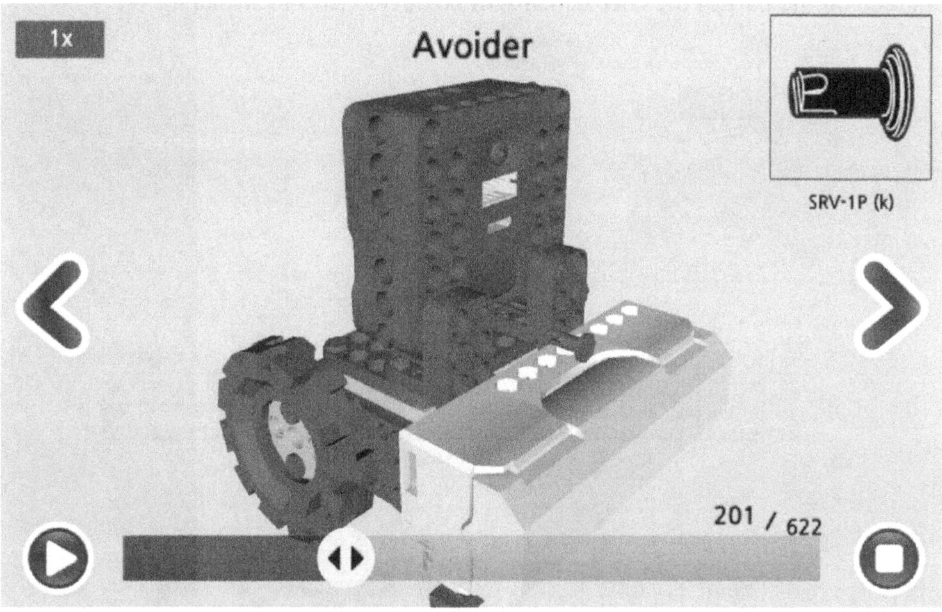

Fig. 2.7 R+m.DESIGN at STEP 201 for Avoider robot.

R+m.TASK's mobile interface is also different from its PC counterpart (see Fig. 2.8), but its overall functionality is the same (see Part 1 of Video 2.7).

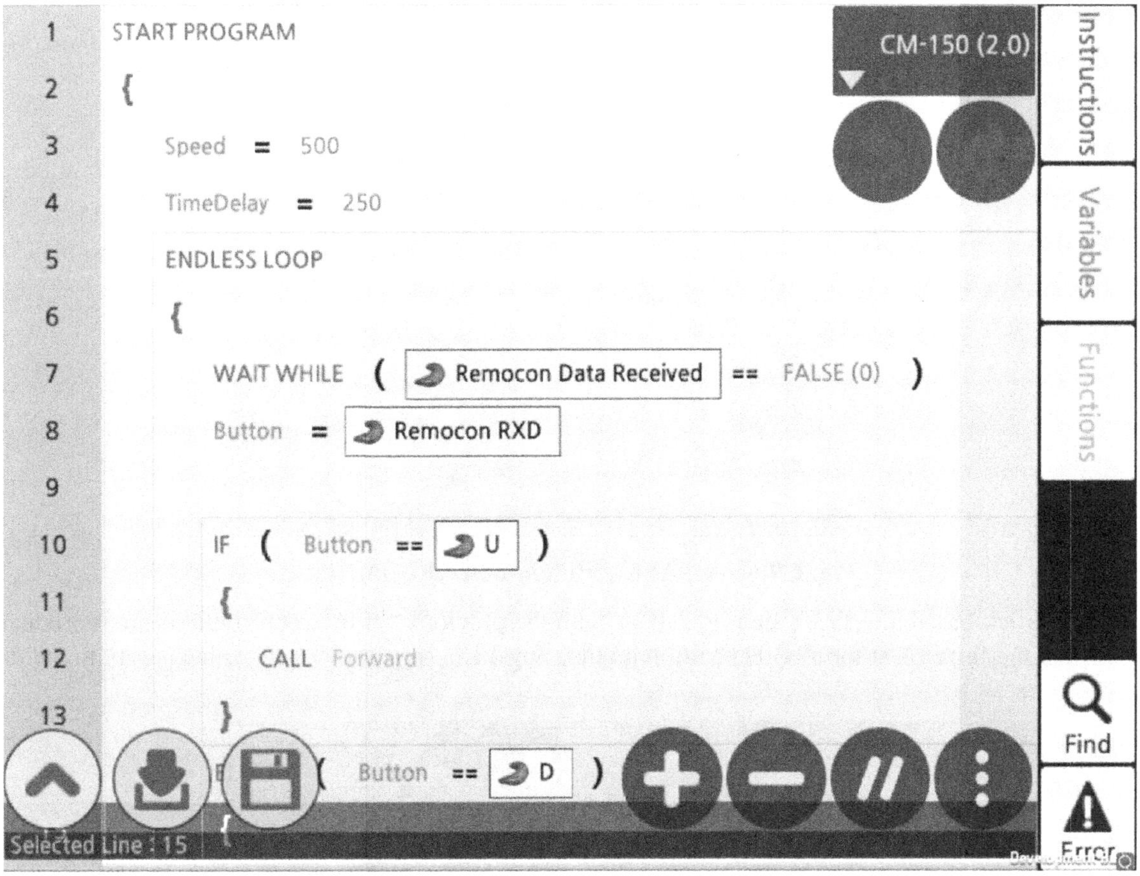

Fig. 2.8 R+m.TASK's Editing Interface.

However, R+m.TASK offers 3 options for its Virtual Controller as compared to only 1 offered on the PC (see Fig. 1.6):

1) A Button-based interface (see Fig. 2.9) where the reader should notice the "greyed out" "in-between" buttons, i.e. between the standard U-D-L-R buttons. For example, the "single" button between "U" and "R" would send the "U+R" value to the robot at run-time. When using the PC-based Virtual Controller, the same effect would require the user to push both "Arrow Up" and "Arrow Right" keys on the keyboard of the PC (see Video 2.3). Furthermore, Buttons 5 and 6 have an "Auto-Pressing" feature, i.e. if the user presses down on them for 2 seconds or longer, they will "lock-in" and stay "pressed down" until the user gives them another quick press to toggle them out of the "Auto-Pressing" mode.

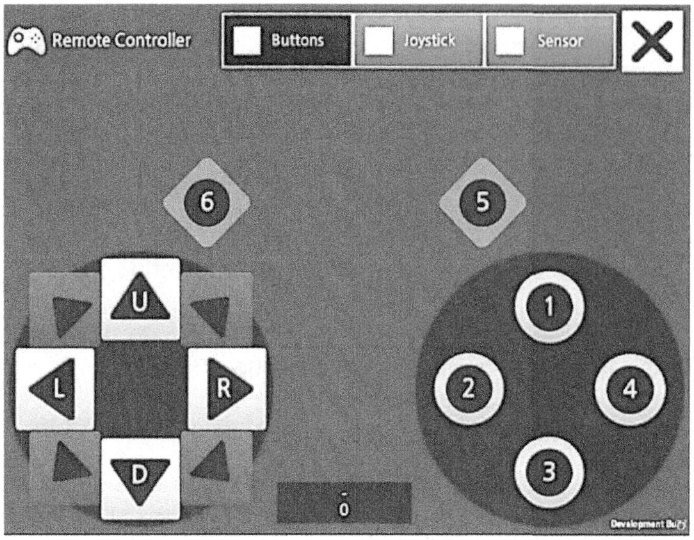

Fig. 2.9 R+m.TASK's Button-based RC Interface.

2) A Joystick-based interface (see Fig. 2.10) where the previous "in-between" buttons are reproduced, i.e. when the finger is at the "2:30" position on the "Joystick" ring, the "U+R" value is sent to the robot at run-time.

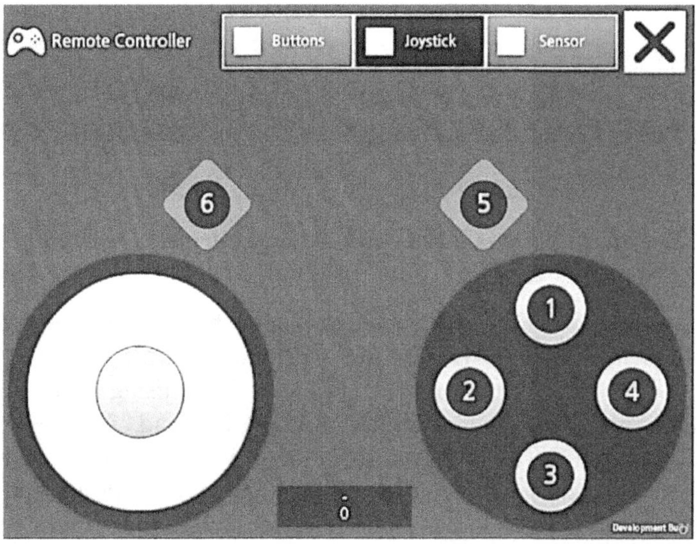

Fig. 2.10 R+m.TASK's Joystick-based RC Interface.

3) A Tilt-Sensor-based interface (see Fig. 2.11) whereas the built-in Tilt Sensor of most mobile devices is used in a similar manner to a physical joystick on an airplane to maneuver the robot. For example, tilting down the front edge of the mobile device, with respect to the horizontal plane, would make the robot go forward, while vice-versa when the front edge is raised up above the horizontal plane. The "in-between" positions similarly to the ones for the Button and Joystick modes are also supported in this Sensor Mode.

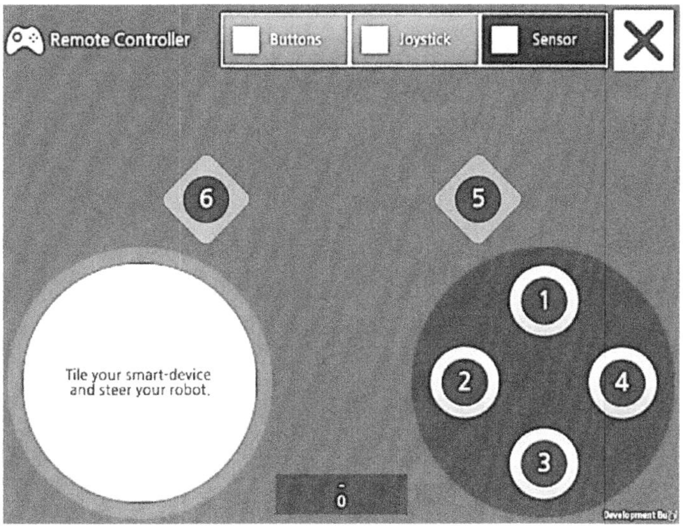

Fig. 2.11 R+m.TASK's Tilt-Sensor-based RC Interface.

These 3 RC tools are showcased in Part 2 of Video 2.7.

2.3 Downloading Example Codes

For PCs and Android devices, the reader needs to first download the Example Codes ZIP file to the Windows PC (see instructions at this web link https://www.cntrobotics.com/dreambook), next unzip it and transfer all the example codes from the PC to the Android device via Windows Explorer (please see Part 1 of Video 2.8). For iOS devices, the procedure is more involved as the Windows PC normally can only "see" the Photo Gallery on these devices. The author uses the "iMazing" software which can transfer seamlessly the example codes from the PC over to the iOS device's file directory (see Part 2 of Video 2.8). The author has not done a thorough Web search to see if there are some "free" software or App that can do similar files transfer tasks as with iMazing.

2.4 References

Thai CN (2018) Advanced Application Programming with ROBOTIS DREAM System. Amazon Kindle Direct Publishing.

2.5 Review Questions for Chapter 2

1. List the memory address for the Center IR Sensor on the CM-150 Controller.
2. List the memory address for the Port 1 Motor on the CM-150 Controller.
3. List the types of "EXECUTE" commands available in the TASK tool.
4. List the types of "LOOP" structures available in the TASK tool.
5. List the types of ROBOTIS Controllers that are supported by the R+SCRATCH App.
6. How many versions of the Virtual Controller are available on the R+m.TASK interface?

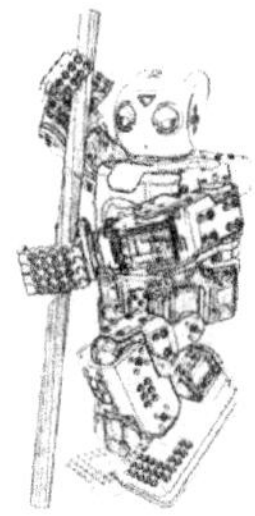

Chapter 3: Using R+TASK & R+m.TASK

This chapter is written for a complete beginner in robotics and computer programming, and we will be using the "Basic Avoider" platform and its variance (Fig. 2.1) to learn how to use the TASK tool. A "Spiral" learning approach (Bergin, 2012) will be used for instructional purposes where the author would show how to use the TASK tool with projects having increasing scope and difficulty.

The TASK tool offers most of the standard programming structures found in other languages such as C/C++. Below is a short description of the most useful features for a beginning programmer (more details on specific items will be provided at appropriate sections in this chapter):

1) ASSIGNMENTS – TASK provides two types of assignment statements: LOAD and COMPUTE.

The syntax for LOAD is "A = B" with the usual meaning of "Operand A is assigned the Value of Operand B".

The syntax for COMPUTE is "A = B *Op* C" where the RHS represents the Value obtained when performing the "*Op*" operation between Operands B and C. The "*Op*" operations supported are the 5 basic arithmetic operators "+, - , * , / , and % (remainder operation)", and the 2 bit-level operators such as "&" (AND) and "|" (OR). Only INTEGER arithmetic is supported. VARIABLE parameters are supported but arrays are not.

2) LABEL and JUMP statements are supported.

3) CONDITIONS – For Conditional statements, TASK provides the usual logical operators: "&&" and "||" respectively for logical AND and logical OR operations, and also "==", "!=" , "<", "<=", ">", ">=" for equality and inequality tests. The standard IF, IF-ELSE-IF, ELSE structures are supported, but nested conditional statements using parentheses are not supported (so the user will have to apply DeMorgan's theorems to expand complex logical statements as needed).

4) LOOPS – for Repetition statements, TASK provides "WAIT WHILE", "LOOP WHILE", "LOOP FOR", "ENDLESS LOOP" and "BREAK LOOP".

5) FUNCTION definition and calling are provided in TASK. A special CALLBACK function can also be used: it is triggered by a hardware timer every 8 ms independently of the main TASK program.

3.1 Learning Sequence Control

Please review Video 2.3 or Video 2.7 if you had not done so to refresh your memory about navigating the menus of the TASK tool on a PC or mobile devices.

In this unit, we are going to learn that by default the CM-150 controller behaves sequentially, i.e. it can do only one thing at a time. Video 3.1 shows how to create a TASK program from scratch, please save this program as "SequenceControl-1.tskx". Video 3.1 also shows how this work was done on PCs and on mobile devices (Android and iOS). Alternately, the reader can use the TASK tool to open the same pre-written program from the provided ZIP file containing all the example codes for this book (see Section 2.3). Fig. 3.1 is a screen capture of this program which has 5 lines of "relevant" code (lines 4-8). More formally, they are called "Statements". All TASK program begins with the "START PROGRAM" statement (line 2) and an opening curly bracket "{" statement (line 3), and ends with a closing curly bracket "}" statement (line 9).

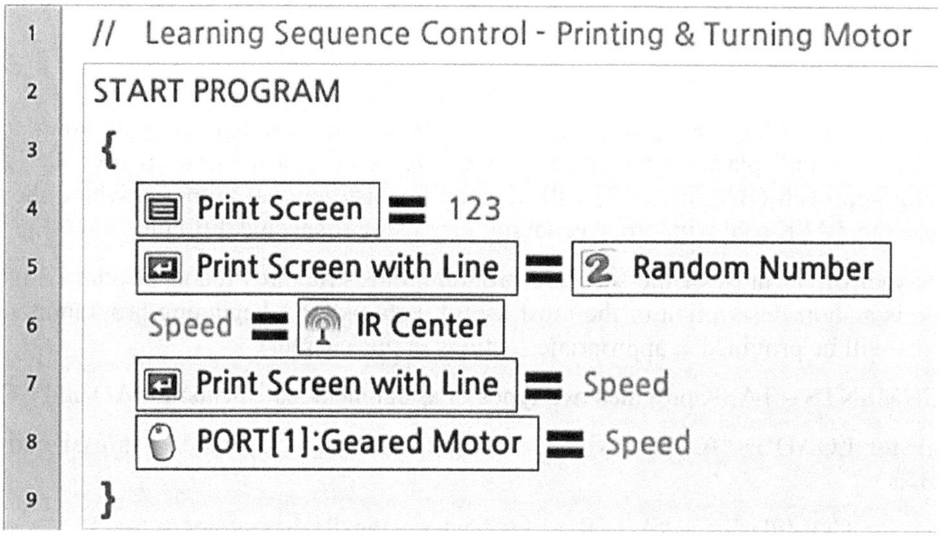

Fig. 3.1 Program "SequenceControl-1.tskx".

All statements between the two matching curly brackets "{}" are considered "legitimate" instructions (but may not be "logically correct") for the CM-150 controller to execute in sequence, top to bottom, at run time. But if they are preceded by the double forward slashes, "//", then they are considered as "Comments", i.e. information for the human reader only and not actual instructions for the controller. For example, Line 1 just tells the reader about the goals of this program. Later, the reader will be shown that it is sometimes necessary to "nest" the curly brackets to create "compound statements" to accomplish certain logical flows needed in a wanted algorithm.

Statements 4 to 8 are called LOAD statements which have this form (i.e. syntax): "A = B". The "=" sign is not the regular "equality" sign that the reader may have seen in algebra. In the world of computer programming, the "=" symbol stands for an "Assignment" operator. Each of the letters/symbols "A" and "B" is called "Operand". Thus "A=B" should be read as ("A" is assigned with the Value of "B"), or it can also be read as ("Value of B" is assigned to "A"). Essentially, "B" should be a Data Value from a Numerical Constant (e.g. "123" in Statement 4), or a Source of Data Input (e.g. a Random Number generated by the Controller internally in Statement 5, or a Value read from a Sensor such as IR Center in Statement 6).

Statement 6 also illustrates the application of a user-defined Named Variable called "Speed" which was first used as an Operand A. Afterwards, "Speed", to be exact its numerical value, can be used as a type of Operand B in Statements 7 and 8.

In other words, the beginner programmer should get used to "read" a LOAD statement from the Right-Hand-Side (RHS) to the Left-Hand-Side (LHS) (which may be "unusual" to English readers). As a side note, TASK supports a second Assignment type of statement called COMPUTE (see its use in a later paragraph) and practically 90% of all the Statements in a typical TASK program are of the "Assignment" type.

When the controller executes Line 4 at run time, it would just "assign" its RHS (e.g. the number "123") to the LHS which is the Screen Printer. The user can then see "123" printed out on the Output Screen of the

PC or mobile device, but the display cursor will stay on the same output line afterwards, because the wanted command is a "Print Screen".

Next, when the controller executes Line 5 at run time, it will want to execute the RHS first, i.e. it will go to a specific memory location (Address 77 to be exact – and you may remember this from the previous Video 2.2) and it will pull out whatever random number between 0 and 255 that it can find there. Next, it will "assign" this found value to the LHS which is the Screen Printer. The user can then see this random number printed out on the Output Screen of the PC or mobile device (after the previous "123"), but then it will execute a "carriage return" to go to a new line afterwards, because the wanted command is now "Print Screen with (new) Line".

Next, Statement 6 is another Assignment statement where the current value of the IR Center Sensor (Address 95) is assigned to a Variable named "Speed".

Next, Statements 7 and 8 both use the current Value of Variable "Speed" either to print it out (Statement 7) or to assign it as a speed value to the Port 1 Geared Motor (essentially to turn it on if "Speed" happens to be non-zero).

Please see Video 3.2 for more details on how this TASK program "SequenceControl-1.tskx" would actually behave at run time on a Windows PC using the micro USB cable and then with the combination BT-410 Dongle and Receiver. Video 3.2 also shows the run-time performance of this program with a Tablet whereas the author uses the same BT-410 Receiver on the robot and the built-in BlueTooth Master from the mobile device.

Video 3.2 shows the user that because of all the time delays needed for Bluetooth connections to work, the run time behavior of this TASK program was far from satisfactory off the Tablet. On the PC runs, we saw the display of "123" and the random number, but we did not see the Port 1 Motor turn itself on during run time when "Speed" had a non-zero value. On mobile devices, we could not see any "Print Screen" output at all sometimes. The "programming logic" used was correct as the controller could compile and execute its machine code perfectly fine on either programming platform, so what was going on with our robot or setup? **The key understanding here is that everything in real life, and therefore in robotics also, "takes time" to get done, so we'll have to figure out a way for the controller not to "rush through" its program, so that we, slow-poke humans, can "see" what is going on.** So, one way to solve these problems is to make the robot repeats this sequence of actions "endlessly" as suggested by the Sense-Think-Act paradigm (see Fig. 1.12).

The "SequenceControl-2.tskx" program just puts the actions sequence of "SequenceControl-1.tskx" inside an "Endless Loop" and with an additional feature of an incremental Counter which is started at "0" outside the Endless Loop and incremented by "1" every time a loop gets done. Fig. 3.2 shows this completed program and Video 3.3 shows how it can be created by modifying the previous program "SequenceControl-1.tskx".

In "SequenceControl-2.tskx", besides the new "Endless Loop" structure used (Statements 6, 7 and 15, and a nested set of "{}"), another new computing concept of an incremental counter is also illustrated:

a) First, Variable "Counter" needs to be initiated to zero, before entering the Endless Loop. This action can be considered as part an "Init Event", previously mentioned in Section 1.2 and Fig. 1.16.

b) Next, Variable "Counter" needs to be updated for each loop executed, by using a COMPUTE statement to make this Variable "self-add" "1" each time the Endless Loop is executed at run time (see Statement 13). The COMPUTE statement is of the form "A = B *Op* C" which has the standard Assignment operator " = " separating the RHS expression (i.e. "B *Op* C") and LHS (i.e. "A"). The RHS now has two Operands "B" and "C" linked by a second operator *Op* which can be of 7 types:

" + " standing for the arithmetic addition operation (used in line 13).

" - " standing for the arithmetic subtraction operation.

" * " standing for the arithmetic multiplication operation.

" / " standing for the arithmetic division operation.

" % " standing for the remainder operation (i.e. yields the "remainder" when doing "manual" integer division.

" & " standing for a bit-wise logical AND operation.

" | " standing for a bit-wise logical OR operation.

c) Statement 14 is used to appraise the human user on the current value of the Variable "Counter".

```
1  // Learning Sequence Control - Printing & Turning Motor
2  // With Endless Loop & Incremental Counter
3  START PROGRAM
4  {
5      Counter = 0
6      ENDLESS LOOP
7      {
8          Print Screen = 456
9          Print Screen with Line = Random Number
10         Speed = IR Center
11         Print Screen with Line = Speed
12         PORT[1]:Geared Motor = Speed
13         Counter = Counter + 1
14         Print Screen with Line = Counter
15     }
16 }
```

Fig. 3.2 Program "SequenceControl-2.tskx".

Video 3.4 showed that thanks to the Endless Loop, the "Speed" command to the Port 1 Motor (Statement 12) was repeatedly applied to this motor so that we could now see the resulting physical effects of its turning. Essentially, Statement 8 in Program "SequenceControl-1.tskx" (Fig. 3.1) **did** turn on Motor 1, but only for a few microseconds and then the whole robot program terminated, so we could not observe this phenomenon as it was too fast for us to see, actually too short to overcome the friction inside the motor and the mechanical inertia of the robot.

However, ones could see that the visual feedback of the printed output was still too fast for us humans to handle, thus we need to introduce some time delays in the whole process with the use of the Hi-Resolution

Timer feature of the CM-150 in the next program "SequenceControl-3.tskx" (see Fig. 3.3 and Video 3.5). The Hi-Res Timer is set at Line 12 to 1.0 sec (or 1000 ms). It functions similarly to a stop-watch or a timer on a microwave oven. To use the Hi-Res Timer, we need first to load the Hi-Res Timer Counter with the wanted time delay (1 second in line 12 – remember Address 74 from Video 2.2?). The Hi-Res Timer Counter would then automatically start to count down to zero with a 1 ms decrement. Next, we need to use a WAIT WHILE statement such as in line 13 to create user-defined time slots associated with certain events. The WAIT WHILE statement is "unusual" in the way that it keeps on executing itself, i.e. "does nothing", as long as the CONDITION specified within the two parentheses "()" remains TRUE, for example (Hi-Res Timer Counter > 0.000 sec) in Line 13. Let's now do a "thought experiment" to see how Lines 12 and 13 can work together at run-time:

When Line 12 is executed, the Hi-Res Timer counter is loaded with 1 second = 1000 ms. Next Line 13 is executed, i.e. the CM-150 is commanded to go to the Hi-Res Timer Counter and retrieve its current value which would be a few milliseconds less than 1000 (let's say 990), as some actual milliseconds have elapsed since the execution of line 12. Thus, the condition (Hi-Res Timer Counter > 0.000 sec) now becomes (990 > 0) and would evaluate to TRUE, therefore the CM-150 would be required to execute line 13 once more. So, this cycle of actions would continue for a while, but you can see that eventually this counter would be at 0. At this point, the condition (Hi-Res Timer Counter > 0) evaluates to FALSE (as "0 is not greater than 0") and the CM-150 can now quit Line 13 and then redo another loop restarting at Line 7.

```
1  // Learning Sequence Control - Printing & Turning Motor
2  // With Endless Loop & 1 second delay
3  START PROGRAM
4  {
5      ENDLESS LOOP
6      {
7          Print Screen = 789
8          Print Screen with Line = Random Number
9          Speed = IR Center
10         Print Screen with Line = Speed
11         PORT[1]:Geared Motor = Speed
12         High-resolution Timer = 1.000sec
13         WAIT WHILE ( High-resolution Timer > 0.000sec )
14     }
15 }
```

Fig. 3.3 Program "SequenceControl-3.tskx".

Video 3.5 shows how this program performed at run-time on a PC and a Tablet. This video also showed that the IR Center Sensor maximum values were mostly in the 500's range and that the motor was turning at a rather slow speed. This was because the typical Geared Motor can be set to a maximum value of 1023 (i.e. 100% power) and here it was set with values in the 500's (i.e. at 50% power). Thus the program "SequenceControl-4.tskx" was created with a small change in Statement 11 whereas the value of Speed was multiplied by a factor of 4 before assigning this result as the actual speed value for Port 1 Motor – i.e. using a COMPUTE statement instead of a LOAD statement (see Fig. 3.4). See Video 3.6 for the run-time performance of program "SequenceControl-4.tskx".

```
//  Learning Sequence Control - Printing & Turning Motor
//   With Endless Loop & Higher Speed
START PROGRAM
{
    ENDLESS LOOP
    {
        Print Screen = 1111
        Print Screen with Line = Random Number
        Speed = IR Center
        Print Screen with Line = Speed
        PORT[1]:Geared Motor = Speed * 4
        High-resolution Timer = 1.000sec
        WAIT WHILE ( High-resolution Timer > 0.000sec )
    }
}
```

Fig. 3.4 Program "SequenceControl-4.tskx".

3.2 Learning Program Modularization with Functions

Let's start with the "Maneuvers-1.tskx" program (Fig. 3.5) which uses previous concepts already explained in Section 3.1. But this time, they are combined to make the robot go forward at Speed = 512 for 0.5 second, and a COMPUTE type of statement is used to set separately the DIRECTION and SPEED for each of the Motor Ports, 2 (i.e. right) and 1 (i.e. left), in lines 8 and 9 in Fig. 3.5.

```
1  // Creating Maneuvers
2  // Maneuver 1 - Going Forward
3  START PROGRAM
4  {
5      Speed = 512
6      DelayTime = 0.500sec
7
8      PORT[2]:Geared Motor = CW:0 (0.00%) + Speed
9      PORT[1]:Geared Motor = CCW:0 (0.00%) + Speed
10     High-resolution Timer = DelayTime
11     WAIT WHILE ( High-resolution Timer > 0.000sec )
12 }
```

Fig. 3.5 Program "Maneuvers-1.tskx".

A "COMPUTE" statement is used here (e.g. line 8 of Fig. 3.5) instead of a "LOAD" statement (like in line 11 of Fig. 3.3) because SPEED is a Variable Parameter, currently set to 512 but it can be changed to something else in other parts of the computer program (to be shown later in this Section 3.2). If the need was only for a fixed numerical constant that would not change throughout the robot program, then a "LOAD" statement would be more efficient to use. Video 3.7 shows how to edit COMPUTE statements inside the program "Maneuvers-1.tskx", how this program performed at run-time and how to modify it into "Maneuvers-2.tskx" which used a new type of computer programming structure called "FUNCTION" (see Fig. 3.6).

```
START PROGRAM
{
    Speed = 512
    DelayTime = 0.500sec

    PORT[2]:Geared Motor = CCW:0 (0.00%) + Speed
    PORT[1]:Geared Motor = CW:0 (0.00%) + Speed
    High-resolution Timer = DelayTime
    WAIT WHILE ( High-resolution Timer > 0.000sec )
}

FUNCTION Forward
{
    PORT[2]:Geared Motor = CW:0 (0.00%) + Speed
    PORT[1]:Geared Motor = CCW:0 (0.00%) + Speed
    RETURN
}
```

Fig. 3.6 Program "Maneuvers-2.tskx".

The later part of Video 3.7 showed how to use the MAIN PROGRAM area as a "sandbox" area where the user can try out different coding ideas, while the FUNCTION areas were used as a repository of coding ideas that had worked well. **This is a "shallow" understanding/application of the FUNCTION structure which serves as a useful but intermediary step towards a more complete understanding of the issues later.**

Once the basic Functions "Forward", "Backward". "Right" and "Left" are defined, the program "Maneuvers-3.tskx" shows how to use them to create a sequence of robot maneuvers, perhaps needed to negotiate a maze, e.g. going forward, turn left then going forward again (see Fig. 3.7). These tasks were done by CALLing those predefined Functions "Forward" and "Left" as needed (lines 9, 13 and 17).

Part 1 of Video 3.8 explains the details of how the controller jumped from a CALL statement in the Main program (such as Line 9 in Fig. 3.7) into the Function definition area, executed the statements defined in this Function, then returned to the Main program to execute the next statement (e.g. Line 10).

```
4  START PROGRAM
5  {
6      Speed = 512
7      DelayTime = 0.500sec
8
9      CALL Forward
10     [High-resolution Timer] = DelayTime
11     WAIT WHILE ( [High-resolution Timer] > 0.000sec )
12
13     CALL Left
14     [High-resolution Timer] = DelayTime
15     WAIT WHILE ( [High-resolution Timer] > 0.000sec )
16
17     CALL Forward
18     [High-resolution Timer] = DelayTime
19     WAIT WHILE ( [High-resolution Timer] > 0.000sec )
20 }
```

Fig. 3.7 Program "Maneuvers-3.tskx".

The reader can see that by using Functions the logic flow in the Main program is easier to understand as it is presented in a more modular fashion, perhaps "logically coarser", while pushing the "gory" operational details into the Function definitions themselves. For example, this CM-150 robot has only two motors, but let's say that you buy 2 more motors so your robot now has four motors to control instead of two, and you still need to perform the same "mission" (forward/left/forward). In this case, you would not have to modify the main program at all, but you will need to add new statements to take care of Motors 3 and 4 inside the appropriate Functions "Forward", "Backward", "Left" and "Right" (see Part 2 of Video 3.8).

Another useful application of Functions is to reduce the number of groups of repeating statements in the Main program. Fig. 3.7 shows three groups of the "same" statements involving the Hi-Res Timer (e.g. lines 10-11, 14-15 and 18-19). The program "Maneuvers-4.tskx" shows how to define the Function DELAY (Fig. 3.8) just one time and how to CALL it in the Main program as many times as needed (Fig. 3.9). "Maneuvers-4.tskx" also shows how to use the Variable Parameters Speed and DelayTime to create maneuvers with varying speeds and delay times, i.e. a more adaptive algorithm to help deal with complex mazes.

```
FUNCTION Delay
{
    High-resolution Timer = DelayTime
    WAIT WHILE ( High-resolution Timer > 0.000sec )
    RETURN
}

FUNCTION Stop
{
    PORT[1]:Geared Motor = CCW:0 (0.00%) + 0
    PORT[2]:Geared Motor = CCW:0 (0.00%) + 0
    RETURN
}
```

Fig. 3.8 Delay and Stop Functions in Program “Maneuvers-4.tskx”.

```
// Going Forward for 0.5 s at Speed 512
Speed = 512
CALL Forward
DelayTime = 0.500sec
CALL Delay

// Turn right for 0.25 s at Speed 256
Speed = 256
CALL Right
DelayTime = 0.250sec
CALL Delay

// Go Backward for 0.35 s at Speed 400
Speed = 400
CALL Backward
DelayTime = 0.350sec
CALL Delay

// Stop robot before ending program
CALL Stop
```

Fig. 3.9 Main Part of Program “Maneuvers-4.tskx”.

Fig. 3.9 also shows a good coding practice of setting the value for a parameter needed for a given function, just before calling that function: for example, set up Speed (line 8) before calling Function Forward (line 9). Video 3.9 shows the runtime performance of the program "Maneuvers-4.tskx".

In this Section 3.2, the robot's maneuvers were "timed maneuvers", thus they may not work consistently for different/changing environments such as different ground surfaces that the robot may run on, or for different charge levels of its batteries (low or high). In later sections, we'll use more "conditional" programming techniques so that the same robot code can work in different environments than the one that the code was originally created in.

3.3 "Spinning Top" Maneuvers using IR Center Sensor

We are now ready for our first implementation of the Sense-Think-Act paradigm as previously discussed in Chapter 1. The sensing component will be the IR Center Sensor of the "Spinning Top" robot as shown in Fig. 3.10 (which is physically the same as the Basic Avoider robot).

Fig. 3.10 "Spinning Top" robot and its IR Center Sensor.

The tasks assigned to this robot are as follows:

1) At start, the robot plays a musical melody to let the user know that it is "ready for action".

2) Next, the behaviors of the robot depend on the position of an object relative to the IR Center Sensor, thus the condition-action list to be considered is as follows:

a) If the IR Center Sensor reading is less than 50, the robot stays still.

b) If the IR Center Sensor reading is between 50 and 300, the robot spins right with its speed proportional to the IR Center Sensor reading and it plays the musical scale Do# for 0.1 second.

c) If the IR Center Sensor reading is between 300 and 500, the robot spins left with its speed proportional to the IR Center Sensor reading and it plays a Re# for 0.1 second.

d) If the IR Center Sensor reading is larger than 500, the robot rolls forward at Speed = 1023 to escape from the object and it plays a Fa# for 0.5 second.

This condition-action list indicates that we are dealing with a situation whereas multiple possible events/conditions may occur to the robot, but that one and only one condition can happen at any one time during the runtime performance of this robot.

Using the built-in musical resources of the CM-150, this challenge will be solved in two ways:

1) Using a "Parallel" IF structure (Section 3.3.1).

2) Using an "IF-ELSE-IF" structure (Section 3.3.2).

3.3.1 Solution using Parallel IF Structures

This type of solution calls for a direct "translation" of each item from the considered condition-action list into its own IF compound statement in the TASK code. The program "IR-based-Maneuvers-IFs.tsk" describes this solution approach:

1) The Initial Event is solved by playing Melody 2 out of the CM-150 speaker (see Fig. 3.11). First, "Melody Time" is assigned to the "Buzzer Timer" (line 5), next the "Buzzer Index" is assigned with the value "Melody 2" (line 6). Then a WAIT WHILE statement is used to let the melody plays itself out (line 7). Lastly, Parameter Speed is initialized to zero (line 8).

```
4   // Init Action
5   Buzzer Timer = Melody Time
6   Buzzer Index = Melody No.2 (2)
7   WAIT WHILE ( Buzzer Timer > 0.0sec )
8   Speed = 0
```

Fig. 3.11 Initial Event for program "IR-based-Maneuvers-IFs.tskx".

2) This solution's algorithm (Figs. 3.12 and 3.13) uses an outer endless loop to represent the World-Robot Interactions Cycle (previously depicted in Fig. 1.12). This loop contains a set of "Parallel" IF structures representing each separate condition-action/input-output item from the complete listing as detailed out previously.

3) During each loop, the controller would evaluate/execute this "Parallel IFs" structure in the following manner:

- It acquires the latest IR Center Sensor reading and saves it in Parameter IR_Sensor (line 13 in Fig. 3.12).

- Next, it enters the "Parallel IFs" structure and checks on the first IF statement (line 14 in Fig. 3.12), i.e. the condition (IR_Sensor < 50). If this condition is FALSE, i.e. (IR_Sensor >= 50), the controller goes on to check on the next IF statement (line 18). But if this condition (IR_Sensor <

50) is TRUE, the Function Stop is called (i.e. the robot stays still) and then the controller will go on to check on the next IF statement at Line 18 of Fig. 3.12 any way (even though logically it does not have to). The astute reader would detect "logical inefficiencies" here already, but for now let's just go on with the rest of the algorithm.

```
11 ENDLESS LOOP
12 {
13     IR_Sensor = IR Center
14     IF ( IR_Sensor < 50 )
15     {
16         CALL Stop
17     }
18     IF ( IR_Sensor >= 50 && IR_Sensor <= 300 )
19     {
20         Speed = IR_Sensor
21         CALL Right
22         Buzzer Timer = 0.1sec
23         Buzzer Index = Do# (4)
24         WAIT WHILE ( Buzzer Timer > 0.0sec )
25     }
```

Fig. 3.12 Part 1 of Main Algorithm used in program "IR-based-Maneuvers-IFs.tskx".

• Line 18 checks to see if IR_Sensor is between 50 and 300, if this is TRUE, it makes the robot turn right and play Do# for 0.1 second (Lines 20-24), then the algorithm goes on to check on Line 26 of Fig. 3.13. If this condition (Line 18) turns out to be FALSE, the algorithm would also go to Line 26 next.

```
26     IF ( IR_Sensor >= 300 && IR_Sensor <= 500 )
27     {
28         Speed = IR_Sensor
29         CALL Left
30         Buzzer Timer = 0.1sec
31         Buzzer Index = Re# (6)
32         WAIT WHILE ( Buzzer Timer > 0.0sec )
33     }
34     IF ( IR_Sensor > 500 )
35     {
36         Speed = 1023
37         CALL Forward
38         Buzzer Timer = 0.5sec
39         Buzzer Index = Fa# (9)
40         WAIT WHILE ( Buzzer Timer > 0.0sec )
41     }
42 }
```

Fig. 3.13 Part 2 of Main Algorithm used in program "IR-based-Maneuvers-IFs.tskx".

- Similarly, Line 26 checks to see if IR_Sensor is between 300 and 500, if this is TRUE, it makes the robot turn left and play Re# for 0.1 second (Lines 30-32), then the algorithm goes on to check on Line 34 of Fig. 3.13. If this condition (Line 26) turns out to be FALSE, the algorithm would also go to Line 34 next.

- Finally, at Line 34, the algorithm checks to see if (IR_Sensor > 500), if this turns out to be TRUE, the robot is commanded to run forward at full speed = 1023 (essentially to get away from the user/obstacle), and to play Fa# for 0.5 second. Then it will repeat the Endless Loop one more time at Line 13 with a fresh IR_Sensor value. If this condition (Line 34) is FALSE, then the algorithm also loops back to Line 13.

Essentially, with this solution approach, **all conditions** are tested at run time, even though logically **one and only one** of them can be TRUE at any one time. In any case, Video 3.10 shows that overall this algorithm did the job fine. The reader would also readily notice that when IR_sensor was between 50 and 300, we could hear the Do# note being played out but the robot was not turning right as it was supposed to. This was because the Speed value was set equal to IR_Sensor in the algorithm, and these specific IR_Sensor/Speed values happened to be too low to overcome the friction and inertia inherent in this robot design. However, when the IR_Sensor reached into the range above 300, it could turn left or run away from the user fine.

3.3.2 Solution using IF-ELSE-IF Structures

The IF-ELSE-IF structure is best suited for a situation where logically the robot can only choose **one and only one** alternative condition/action route **among several** potential condition/action routes (which is the current case).

The program "IR-based-Maneuvers-IF-ELSE-IF.tskx" describes the solution approach used:

1) The Initial Event is solved by playing Melody 2 out of the CM-50 speaker (same as in Fig. 3.11). First, "Melody Time" is assigned to the "Buzzer Timer" (line 5), next the "Buzzer Index" is assigned with the value "Melody 2" (Line 6). Then a WAIT WHILE statement is used to let the melody plays itself out (Line 7). Lastly, Parameter Speed is initialized to zero (Line 8).

2) This solution's algorithm (Fig. 3.14) also uses an outer Endless Loop to represent the World-Robot Interactions Cycle of Fig. 1.12. But this loop now contains an IF-ELSE-IF structure representing the condition-action/input-output listing as detailed out previously.

3) During each loop, the controller would evaluate/execute this IF-ELSE-IF structure in the following manner:

- It acquires the latest IR Center Sensor reading and saves it in Parameter IR_Sensor (Line 13).

- Next, it enters the IF-ELSE-IF structure and checks the first IF statement (Line 14), i.e. the condition (IR_Sensor <= 50). If this condition is TRUE, the Function Stop is called (i.e. the robot stays still) and the controller will exit the whole IF-ELSE-IF structure at Line 41 and then it will perform another loop from Line 13. But if this condition is FALSE, i.e. (IR_Sensor > 50), the controller goes on to check the next ELSE-IF statement (Line 18).

```
11  ENDLESS LOOP
12  {
13      IR_Sensor = IR Center
14      IF ( IR_Sensor <= 50 )
15      {
16          CALL Stop
17      }
18      ELSE IF ( IR_Sensor <= 300 )
19      {
20          Speed = IR_Sensor
21          CALL Right
22          Buzzer Timer = 0.1sec
23          Buzzer Index = Do# (4)
24          WAIT WHILE ( Buzzer Timer > 0.0sec )
25      }
26      ELSE IF ( IR_Sensor <= 500 )
27      {
28          Speed = IR_Sensor
29          CALL Left
30          Buzzer Timer = 0.1sec
31          Buzzer Index = Re# (6)
32          WAIT WHILE ( Buzzer Timer > 0.0sec )
33      }
34      ELSE IF ( IR_Sensor > 500 )
35      {
36          Speed = 512
37          CALL Forward
38          Buzzer Timer = 0.5sec
39          Buzzer Index = Fa# (9)
40          WAIT WHILE ( Buzzer Timer > 0.0sec )
41      }
42  }
```

Fig. 3.14 Main Algorithm used in program "IR-based-Maneuvers-IF-ELSE-IF.tskx".

• Please note that according to the considered condition-action listing, the controller is supposed to check if IR_Sensor is between 50 and 300 at Line 18, but it now uses a condition that involves only checking for (IR_Sensor <= 300). Is that sufficient to do what is required? And the answer is that it is OK to just check for the condition (IR_Sensor <= 300) here, because the condition (IR_Sensor > 50) was implicitly TRUE when Line 18 was chosen (see previous paragraph). Thus if (IR_Sensor <= 300) turns out to be TRUE (meaning that IR_Sensor is between 50 and 300), the controller assigns the value of IR_Sensor to Speed and calls the Function Right to make the robot spin right at a Speed proportional to IR_Sensor's value. The controller also plays Do# for 0.1 second, then exits this IF-ELSE-IF structure and makes another loop starting at Line 13. However, if the condition (IR_Sensor <= 300) turns out to be FALSE, i.e. (IR_Sensor > 300), then the controller goes on to execute the next ELSE-IF statement at Line 26.

• Similarly, executing Line 26 means that the controller is checking to see if IR_Sensor is between 300 and 500. If this is TRUE, then the robot spins left with a Speed proportional to IR_Sensor's value, plays Re# for 0.1 second, exits this IF-ELSE-IF structure and continues with its looping task starting at Line 13. If the condition in Line 26 turns out to be FALSE (i.e. (IR_Sensor > 500), then Line 34 is executed next.

• Let's have a pop-quiz for Line 34. **Was it necessary to check for (IR_Sensor > 500) on line 34? Or would a simple ELSE would do the job here?** The author would leave it to the readers to experiment on their own here. If (IR_Sensor > 500) turns out to be TRUE, then the robot would run forward at Speed=512 and for 0.5 second while playing Fa#. This maneuver is designed to make the robot run away from under the object/user. Once away from the object's influence, the IR_Center Sensor would then read zero, and the robot would stop and be ready to "play again" (i.e. Lines 14-17).

Video 3.11 shows the runtime performance of this solution "IR-based-Maneuvers-IF-ELSE-IF.tskx" which was not that different from the solution "IR-based-Maneuvers-IFs.tskx". In theory, the IF-ELSE-IF version is computationally more efficient and therefore "faster" than the Parallel IFs version. In the Exercise Section at the end of this chapter, the reader will have the occasion to use a "timed" version of both solutions to concretely do the comparison (the results will surprise the reader!).

3.4 "Avoider" Maneuvers using Left and Right IR Sensors

In this section, we'll use the same robot platform as shown in Fig 3.10, but it is renamed "Avoider" to emphasize the types of maneuvers described herein. The "Avoider" will be using its Left and Right IR Sensors to detect a single object and to accomplish tasks described in the following condition-action list:

1) **If there is no obstacle in front of Avoider, it can go forward.**

2) **If there is an object on the Left of Avoider, it should back away and turn to the Right to avoid the obstacle.**

3) **If there is an object on the Right of Avoider, it should back away and turn to the Left to avoid the obstacle.**

4) **If there is an object in front of Avoider, it should go straight Back to avoid the obstacle.**

At first analysis, this seems to be a familiar programming situation of choosing one and only one alternative among several possible alternatives, thus an IF-ELSE-IF structure should do an adequate job when combined with an Outer Endless Loop.

The TASK solution is named "Avoider-1.tskx". It starts with the playing of Melody 2 (same code as shown in Fig. 3.11). The main algorithm is described in Fig. 3.15:

1) The Endless Loop starts out by reading in the latest values from the IR Left and Right Sensors and saves them respectively into Variable Parameters IR_Left (Line 14) and IR_Right (Line 15).

2) Next is an IF-ELSE-IF structure with 4 mutually exclusive cases:

• At Line 16, if the values of Parameters "IR_Left" and "IR_Right" are both less than 10 (i.e. no obstacle is in front of the robot), the robot goes forward and the controller would then exit the IF-

ELSE-IF structure and restart another loop at Line 14. If this condition is FALSE, i.e. when either "IR_Left" <u>or</u> "IR_Right" have values larger than 10, or in other words when either IR sensor sees an object, the controller would need to check further by executing the next ELSE-IF statement at Line 20.

```
12  ENDLESS LOOP
13  {
14      IR_Left  =  IR Left
15      IR_Right  =  IR Right
16      IF  (  IR_Left  <  10 && IR_Right  <  10  )
17      {
18          CALL Forward
19      }
20      ELSE IF  (  IR_Left >= 100 && IR_Right  <  10  )
21      {
22          CALL RightBack
23      }
24      ELSE IF  (  IR_Right >= 100 && IR_Left  <  10  )
25      {
26          CALL LeftBack
27      }
28      ELSE IF  (  IR_Left >= 100 && IR_Right >= 100  )
29      {
30          CALL Backward
31      }
32  }
```

Fig. 3.15 Main Algorithm used in program "Avoider-1.tskx".

- The condition ((IR_Left >=100) AND (IR_Right < 10)), used for the ELSE-IF statement at Line 20, represents a situation where there is an object on the left of the robot and none of its right side. If this condition evaluates to TRUE, the robot would turn right by backing away to avoid the obstacle on its left (Line 22). The controller next exits this IF-ELSE-IF structure, and loops again at Line 14. If this condition evaluates to FALSE, then Line 24 is executed next.

- Similarly, the condition ((IR_Right >=100) AND (IR_Left < 10)), used for the ELSE-IF statement at Line 24, represents a situation where there is an object on the right of the robot and none of its left. If this condition evaluates to TRUE, the robot would back away and turn left to avoid the obstacle on its right (Line 26). The controller next exits this IF-ELSE-IF structure, and loops again at Line 14. If this condition evaluates to FALSE, then Line 28 is executed next.

- The ELSE-IF statement at Line 28 uses this condition ((IR_Left >=100) AND (IR_Right >= 100)) to represent a situation when there is an object close enough to the robot's front such that both IR Sensors report data values larger than 100. If this condition is TRUE, the robot would roll straight back away from the obstacle, otherwise the controller loops back to Line 14.

Video 3.12 shows the run-time performance of this algorithm where the robot handled well the situations when the object was on either side of itself. But it did not handle well the situation when a "flat" object was right in front of it, as it kept on going back and forth at the same spot, i.e. it got "stuck". **This is a side-effect of the current algorithm that resulted in the controller "bouncing" back and forth between the conditions represented by Line 16 and Line 28.**

This problem can be easily fixed by adding a turning maneuver to the current backward maneuver (Line 30). The turn direction does not matter, the author happens to use a left turn in the revised solution "Avoider-2.tskx" (see Fig. 3.16, Lines 30-31). Video 3.13 shows much improved performance for frontal objects and corners.

```
28    ELSE IF ( IR_Left >= 100 && IR_Right >= 100 )
29    {
30        CALL Backward
31        CALL LeftBack
32    }
```

Fig. 3.16 Adding Left Turn to Backward Maneuver in program "Avoider-2.tskx".

The next program "Avoider-3.tskx" offers a different solution technique to these maneuvering tasks. **This new approach still relies on the same condition/action listing as described at the beginning of this section 3.4, but it distills different behavior patterns from this list:**

1) We must identify a "default" action which in this case is "Going Forward".

2) We will be using several simple IFs in parallel instead of a single big IF-ELSE-IF structure like before. **We also need to do a deeper analysis of the condition/action list to retain only the "essential" condition/action pairs.**

3) We also **need to apply a Delay Time to each action made** (this part was not needed when the IF-ELSE-IF structure was used). This Delay Time can be tuned for optimal performance.

Fig. 3.17 shows the relevant section of the algorithm used:

- Line 10 showed that a Delay Time of 100 ms was used (the reader should try different values such as 0, 50, 150 etc… to see the effect on the overall performance of the robot).

- An ENDLESS LOOP is still used (Line 13) and starts by updating the parameters IR_Left and IR_Right with the latest data input from the IR Left and Right Sensors (Lines 15 and 16).

- The first IF statement (Lines 17-21) represents the condition when there is an object on the robot's left (Line 17), then the robot would back away and turn right (as before) to steer away from the obstacle (Line 19). A short delay (Line 20) is necessary for the robot to stay in this "action" to make this algorithm work properly.

- The second IF statement (lines 22-26) represents the condition when there is an object on the robot's right (Line 22), then the robot would back away and turn left (as before) to steer away from the obstacle (Line 24). A short delay (Line 25) is also necessary for the robot to stay in this "action" to make this algorithm work properly.

- **Please note that the "default" maneuver is always programmed as the "Last Action" ("Going Forward" in this case – Line 27).**

```
9   Speed = 256
10  DelayTime = 100
11
12  // Main Loop
13  ENDLESS LOOP
14  {
15      IR_Left = IR Left
16      IR_Right = IR Right
17      IF ( IR_Left >= 100 )
18      {
19          CALL RightBack
20          CALL Delay
21      }
22      IF ( IR_Right >= 100 )
23      {
24          CALL LeftBack
25          CALL Delay
26      }
27      CALL Forward
28  }
```

Fig. 3.17 Alternate Maneuvering Algorithm using "Parallel IFs" in program "Avoider-3.tskx".

It is important to perform a "thought experiment" here to "preview" the possible results of this Endless Loop at run time:

- **Every time that the Endless Loop is executed, Line 27 will be executed**, i.e. the robot will always go forward (a little bit), regardless of everything else that may happen.

- Regarding the two parallel IFs, during some execution cycle(s), none of them could be activated (i.e. when there is no obstacle), or one of them could be triggered (i.e. when the obstacle is on either side of robot), or both could be triggered (i.e. a frontal obstacle). The alert reader may have noticed that seemingly the case for "a frontal obstacle" was not dealt with in this algorithm, so how would the robot behave at run time when encountering a frontal obstacle? Watching Video 3.14 will show the reader that the robot "wiggles" left-right and backward to deal with a frontal obstacle, i.e. the Functions RightBack and LeftBack were alternatively called. **This effect also explains why a DelayTime=0 would not work for the robot, as it would be switching so fast between the Functions "RightBack", "LeftBack" and "Forward" (within the Endless Loop) such that the robot looks "paralyzed" from the outside.** By the way, the same thing happens to us humans when we multitask too much as nothing then gets accomplished!

At this point, some of the more advanced readers may have wondered whether the IF statements used in Lines 17 and 22 of Fig. 3.17 could be changed to LOOP WHILE statements? And would the robot still work OK for this challenge?

The program "Avoider-4a.tskx" implements this idea (see Fig. 3.18) and Video 3.15 shows that this idea did not work! What went wrong? Was it wrong to use LOOP WHILE for this challenge? Watching Video 3.15 carefully would help the reader figure the "error" made.

```
ENDLESS LOOP
{
    IR_Left  =  IR Left
    IR_Right  =  IR Right
    LOOP WHILE  (  IR_Left >= 100  )
    {
        CALL  RightBack
        CALL  Delay
    }
    LOOP WHILE  (  IR_Right >= 100  )
    {
        CALL  LeftBack
        CALL  Delay
    }
    CALL  Forward
```

Fig. 3.18 Incorrect use of "LOOP WHILE" in program "Avoider-4a.tskx".

As it turns out, the error was not because LOOP WHILE was used per se, but it was more because LOOP WHILE was implemented incorrectly, as we forgot to "feed" them with the latest IR sensor data. We were using the Variable Parameters IR_Left and IR_Right to control the loops and we forgot to update their values every time we went through the LOOP WHILE (as we were supposed to!). It was a simple change to get to the program "Avoider-4b.tskx" (see Fig. 3.19) which does work properly. Please look at Lines 17 and 22 in Fig. 3.19 where the real-time data from the sensors' ports are now accessed and used to control the LOOP WHILE, not <u>copies</u> of their values, i.e. parameters IR_left and IR_Right, as before. Please also note that lines 15 and 16 are commented out as they are no longer needed. See Video 3.16 for a runtime performance of this algorithm.

```
ENDLESS LOOP
{
    // IR_Left = IR Left
    // IR_Right = IR Right
    LOOP WHILE ( IR Left >= 100 )
    {
        CALL RightBack
        CALL Delay
    }
    LOOP WHILE ( IR Right >= 100 )
    {
        CALL LeftBack
        CALL Delay
    }
    CALL Forward
}
```

Fig. 3.19 Correct Use of "LOOP WHILE" in program "Avoider-4b.tskx".

There is an interesting experiment that the reader can make with the previous different versions of the Avoider codes: create a maze using books or boxes and let the Avoider robot roam freely through it. You will be surprised with its behaviors depending on which TASK code you used, especially when your maze has a cul-de-sac.

3.5 "Follower" Maneuvers using Left and Right IR Sensors

To round out the reader's experience with using the Left and Right IR Sensors, let's apply a **"reverse logic"** on them, essentially let's make the robot follow an object when it detects something in front of it.

Let's first learn how to track a moving object, when it is detected. The corresponding condition/action list can be as follows:

1) **If there is no object detected, the robot should stay still (default behavior).**
2) **If there is an object detected by the Right IR Sensor, the robot should swing right.**
3) **If there is an object detected by the Left IR Sensor, the robot should swing left.**

Should we also plan for the case when an object is detected and located right in front of the robot? Should it "stop" or should it show that it already "locks in" on the object in some other way? From the preceding Avoider coding work, we should be able to let the robot bounce between its two actions of swinging left and right to show that it got the object in its sight, right?

The program "Follower-1.tskx" is derived from the previous discussions using LOOP WHILEs. The algorithm used is shown in Fig. 3.20:

- Essentially, whenever any IR Sensor (Left or Right) picks up on an object (Lines 15 or 19), the robot is turned, respectively, towards the same side where that IR Sensor is located (Lines 17 or 21). Please note that a low detection threshold value of "5" is used so that the robot can detect an object as "far" as possible (about 5 inches away)!

```
13  ENDLESS LOOP
14  {
15      LOOP WHILE ( IR Left >= 5 )
16      {
17          CALL Left
18      }
19      LOOP WHILE ( IR Right >= 5 )
20      {
21          CALL Right
22      }
23      CALL Stop
24  }
```

Fig. 3.20 Object tracking algorithm in program "Follower-1.tskx".

- The default maneuver is at Line 23 which is interestingly a "non-maneuver" as it keeps the robot still (CALL Stop)!

Video 3.17 shows the runtime performance of this algorithm. **From Video 2.2, please remember the broad angle of view supported by the built-in Left and Right NIR Sensors, and please note the effect of the object's size on how the robot would tract the object.**

Here, the reader is challenged to come up with another algorithm using parallel IFs instead, however this new algorithm should yield the same object tracking result. Don't forget to use a small Delay Time to make this approach work!

In the next "Follower" application, a new condition/action pair is added to the previous list:

- **When the object gets a bit closer to the robot (i.e. threshold value of 75 applied to both IR Sensors), the robot should get even closer to the object by going Forward. At this point during runtime, if the user then moves the object away from the robot at a proper speed, ones would see the effect of the robot "following" the object at a short distance away.**

Surprisingly, there are two possible solutions: one using a combination of LOOP WHILE and an IF-ELSE-IF structure (Follower-2.tskx), and another one using only an IF-ELSE-IF structure (Follower-3.tskx). Who knows may be the readers can find other solutions too!

The program "Follower-2.tskx" (see Fig. 3.21) uses a default maneuver (CALL Stop) and an outer LOOP WHILE which encompass an inner IF-ELSE-IF structure that determines the one appropriate action among 3 possible and mutually exclusive alternatives "Left", "Right" or "Forward".

```
13  ENDLESS LOOP
14  {
15      LOOP WHILE ( IR Left >= 5 || IR Right >= 5 )
16      {
17          IF ( IR Left >= 5 && IR Right < 2 )
18          {
19              CALL Left
20          }
21          ELSE IF ( IR Right >= 5 && IR Left < 2 )
22          {
23              CALL Right
24          }
25          ELSE IF ( IR Left >= 75 && IR Right >= 75 )
26          {
27              CALL Forward
28          }
29      }
30      CALL Stop
31  }
```

Fig. 3.21 Object tracking algorithm in program "Follower-2.tskx".

Or in everyday language:

- When no obstacle is detected, the LOOP WHILE (Lines 15-29) would be skipped and Line 30 would be executed, i.e. the robot would stay still as its default behavior.

- When <u>either</u> IR Sensor detects an object (Line 15), the LOOP WHILE is entered and the controller checks further with the IF-ELSE-IF structure to see if the IR Left Sensor is the only one detecting the object (threshold "5" exceeded in Line 17) then the robot goes Left (Line 19). But if "Line 17" turns out to be FALSE, the controller next checks to see if the IR Right Sensor is the only one detecting the object (threshold "5" exceeded in line 21), then the robot goes Right (Line 23). But if "Line 21" turns out to be FALSE also, then the controller next checks to see if the object is close to both IR Sensors (Line 25). If "Line 25" turns out to be TRUE (i.e. both IR Sensors exceed a threshold of "75"), then the robot would go Forward (Line 27). But if "Line 25" turns out to be FALSE, the controller reaches the end of the IF-ELSE-IF structure without being able to determine a concrete action to do, so the controller would just loop back to Line 15 to check on the condition of the LOOP WHILE "((IR_Left >= 5) OR (IR_Right >= 5))". If this condition is still TRUE, it would search through the IF-ELSE-IF structure again. If this condition becomes FALSE, the controller finally quits the LOOP WHILE, next triggers Line 30 to stop the robot and then redo the outermost ENDLESS LOOP at Line 14 all over again.

- Please note, once again, that when using LOOP WHILE, we need to use "current" IR Sensor Data inputs in all our conditional statements, so that most recent data from the sensors are used. Please see Video 3.18 for the runtime performance of "Follower-2.tskx".

The algorithm used in the next program "Follower-3.tskx" is described in Fig 3.22. It is more straight forward so it is easier to explain:

• It uses an outermost ENDLESS LOOP (as always) (Lines 13 to 33) and the first actions by the controller is to go out to the IR Sensors Ports and to read in the latest IR values and save them in parameters IR_Left and IR_Right (Lines 15 and 16).

• Next is the familiar IF-ELSE-IF structure which now represents 4 mutually exclusive condition/action pairs for "Left", "Right", "Forward" or "Stop". Please note that the IF-ELSE-IF structure in the previous "Follower-2.tskx" handles only 3 condition/action pairs.

• If Line 17 is TRUE, i.e. only the IR Left Sensor sees the object, so the robot turns left.

• Otherwise, Line 21 is next executed, and if Line 21 turns out to be TRUE, i.e. only the IR Right Sensor sees the object, then the robot turns right.

• Otherwise, Line 25 is executed next, and if Line 25 turns out to be TRUE, i.e. both IR sensors see the object up close, then the robot goes forward to get even closer to the object.

• Lastly, when no condition turns out to be TRUE (ELSE in Line 29), the robot executes its default behavior, i.e. staying still.

Video 3.19 shows the run-time performance of this program "Follower-3.tskx".

```
13  ENDLESS LOOP
14  {
15      IR_Left  =  IR Left
16      IR_Right  =  IR Right
17      IF ( IR_Left >= 5 && IR_Right < 2 )
18      {
19          CALL Left
20      }
21      ELSE IF ( IR_Right >= 5 && IR_Left < 2 )
22      {
23          CALL Right
24      }
25      ELSE IF ( IR_Left >= 75 && IR_Right >= 75 )
26      {
27          CALL Forward
28      }
29      ELSE
30      {
31          CALL Stop
32      }
33  }
```

Fig. 3.22 Object tracking algorithm in program "Follower-3.tskx".

There was no discernable difference in performance between these two solutions that the author could see. In theory, "Follower-3.tskx" should be "faster" because it accesses the sensor ports only twice per each iteration of the overall Endless Loop, i.e. it uses less resources from the controller as compared to "Follower-2.tskx" which uses from 4 to 8 accesses to the sensor ports per each iteration of its overall Endless Loop. However, "Follower-2.tskx" can be more "accurate" in its actions as it always uses "fresh" data inputs. In other words, there is no "free lunch" in life and in robotics.

Please note that ELSE is used in Line 29 of "Follower-3.tskx", and its use is "critical" here. As an exercise for the reader, try to comment out Line 29 (i.e. to make "Call Stop" the default behavior) and you will find that the robot no longer works properly. But if you **then add proper Delay Times** (~100 ms) for the actions "Left", "Right" and "Forward", you will find that the robot is working properly again! (see "Follower-3a.tskx"). So once again, the robot seems to work better when it is slowed down an appropriate amount!

3.6 Autonomous Behavior Programming Summary

Recapping Sections 3.4 and 3.5, the author's goal was to show that programming "autonomous" behaviors can be fun and challenging at the same time! Starting from the (now familiar) condition/action analysis, **three techniques can be used to create TASK codes and after some initial actions, they all used an outermost ENDLESS LOOP for the main algorithm:**

1) In Technique 1, for each iteration of the Endless Loop, latest readings from the used sensors were first recorded, then an IF-ELSE-IF structure was used to represent all possible but mutually exclusive condition/action pairs that corresponded to robot behaviors required to fulfill a given challenge. Example programs using Technique 1 are "IR_based_Maneuvers-IF-ELSE-IF.tskx", "Avoider-1.tskx", "Avoider-2.tskx", and "Follower-3.tskx".

2) In Technique 2, the sensors reading task is still required first, but the IF-ELSE-IF structure is replaced by "parallel" IFs. This technique requires a deeper analysis of the original condition/action list to identify a "default" behavior, and other condition/action pairs that can work "on their own" or "in concert". This technique also requires a short time delay incorporated into each robot action. Compared to Technique 1, Technique 2 usually results in a smaller number of statements used in the TASK code, meaning that the controller would take less time to do each iteration of the Endless Loop, and thus the robot may become more responsive to changes in the "World" (review Fig. 1.12). Example programs using this technique are "Avoider-3.tskx" and "Follower-1-IFs.tskx".

3) Technique 3 is a variation of Technique 2 and it uses "parallel" LOOP WHILEs instead of "parallel" IFs. It combines "Data Acquisition" tasks in its Conditional Statements, thus it uses the least number of statements, as compared to the other two Techniques. Example programs using Technique 3 are "Avoider-4b.tskx" and "Follower-1.tskx". "Follower-2.tskx" is a hybrid solution combining Techniques 3 and 1.

3.7 Remote Control Programming Techniques

Back in Section 1.1, "Instruction" and "Status" packets were conceptually described, and "RemoCon" packets were first mentioned. In this Section 3.7, a more detailed description of "RemoCon" packets is provided, along with programming examples of different Remote Control schemes for the "Avoider" robot:

1) Remote Control from PC-based TASK tool: Virtual and Physical RC-100B controller.

2) Remote Control from Mobile R+m.TASK tool: the 3 modes of the Mobile Virtual RC-100 Controller.

3.7.1 Structure of the "Remocon" packet

A full "Remocon" packet is based on a 16-bit user-designed message wrapped up in a 6-byte packet (http://support.robotis.com/en/product/auxdevice/communication/rc100_manual.htm). As this book is for younger beginning programmers, we'll restrict ourselves to using only the lower 10 bits of this 16-bit message. Advanced readers should consult Chapter 7 of Thai (2017) for a more complete treatment of this topic.

Both the Physical RC-100B Remote Controller and the multiple Virtual RC-100 Remote Controllers (see Figs 1.5, 1.6, 2.9, 2.10 and 2.11) were designed to behave programmatically in the same manner:

- Each of the 10 buttons U-D-L-R-1-2-3-4-5-6 is represented by 1 bit in a 10-bit number/message, with the "U" button using the bit-0 position and the "6" button using the bit-9 position (see Fig. 3.23).

RC-100 10-bit Message

Bit Position	*9*	*8*	*7*	*6*	*5*	*4*	*3*	*2*	*1*	*0*
Button	6	5	4	3	2	1	R	L	D	U
Example Data	**0**	**0**	**0**	**1**	**0**	**0**	**1**	**0**	**0**	**1**

Fig. 3.23 Structure of an RC-100 10-bit message.

- Whenever one or several buttons are pushed by the user, this 10-bit message is sent from the Physical and Virtual RC-100 to the robot. The "Example Data" row in Fig. 3.23 shows that Buttons "U", "R" and "3" are pushed as these bits are set to "1". All the "un-pushed" buttons of course have their bits set to 0.

- Lastly, whenever the user releases all buttons, the Physical and Virtual RC-100 would send out the number "0" (i.e. all bits set to 0) ONCE. This feature is unique with the ROBOTIS RC-100 system, other manufacturers may not implement the same scheme.

The main issues in using the RC-100 are to know how many buttons are being pushed, and how to sort out which buttons are being pushed, as information about **all 10 buttons** are sent over inside each transmitted RC message.

First, the issue of "how many" is controlled by the programmer, as readers may well know that it is impossible to expect that actual users of one's programs will only push the buttons that the programmer would want them to! The program "RC-1Button.tskx" (Fig. 3.24) shows how to restrict one's program to deal only with 1 "expected" button ("U" or "D") and to ignore any other situation.

After the usual Initial Action section whereas Melody 2 is played and Variables Data and DelayTime are initiated respectively to 0 and 250 (i.e. 250 ms), the main algorithm is shown in Fig. 3.24:

```
12  ENDLESS LOOP
13  {
14      WAIT WHILE ( Remocon Data Received == FALSE (0) )
15      Data = Remocon RXD
16      Print Screen with Line = Data
17      IF ( Data == U )
18      {
19          Print Screen = Data
20          Print Screen with Line = 11
21      }
22      ELSE IF ( Data == 0000 0000 0000 0000 0000 0000 0000 0010 )
23      {
24          Print Screen = Data
25          Print Screen with Line = 22
26      }
27      CALL Delay
28  }
```

Fig. 3.24 Main algorithm in program "RC-1Button.tskx".

- Statements 14 and 15 need to be used together and they represent the first actions to be done in a basic RC Loop. The CM-150 Controller's Firmware has a built-in procedure that keeps on listening on the RC communication channel and if a new message has been received, it sets the "Remocon Data Received" flag to TRUE, otherwise this flag is set to FALSE. Thus, at each iteration of the Endless Loop (Lines 13-28), the CM-150 would first check the WAIT WHILE's condition set in Line 14. If no button has been pushed, this condition evaluates to (FALSE == FALSE), therefore this WAIT WHILE's condition evaluates to TRUE, and the CM-150 would stay at Line 14. Eventually, the user pushes one or many buttons on the RC-100. This would make the flag "Remocon Data Received" become TRUE and the WAIT WHILE's condition now becomes (TRUE == FALSE) which evaluates to FALSE. When this happens, Line 14 is exited and Line 15 is executed next. It goes to the memory buffer "Remocon RXD" and save the value found in there into the Parameter "Data". Line 16 essentially prints out a decimal number that represents the status of all 10 buttons.

- As this program is only interested in the event whereas only ONE button has been pushed and furthermore this single button needs to be either "U" or "D" (but not both), the IF-ELSE-IF structure (Lines 17-26) implements these requirements. Line 17 checks to see if "Data" is equal to Button "U", if this turns out to be TRUE, the CM-150 prints "Data" and then the number "11" as a marker (Lines 19-20). If Line 17 turns out to be FALSE, then Line 22 is executed next. Line 22 checks to see if "Data" is equal to Button "D", here Button "D" is shown as its binary-number equivalent (showing that Bit 1 is set). If Line 22 turns out to be TRUE, the CM-150 prints "Data" and then the number "22" as a marker (lines 24-25). If Line 22 turns out to be FALSE, then Line 27 is executed next (i.e. a 250 ms delay so that the user can reasonably read the Output Window inside the TASK tool). Then another iteration of this Endless Loop begins again at Line 14.

Video 3.20 shows a typical run of this program whereas all messages sent by the RC-100 were shown, but only those single-button ("U" or "D") messages were processed further, i.e. printed out with either marker "11" or "22" as appropriate.

The next program "RC-2Buttons.tskx" shows how to parse out two specific buttons "Button1" and "Button2" from the RC-100's 10-bit message. "Button1" needs to be a single-button input from the group "U-D-L-R", while "Button2" needs to be a single-button input from the group "1-2-3-4". The main algorithm is shown in Fig. 3.25 and it has two new features - a binary filtering procedure using the bit-level "&" operator (Lines 20 and 22) and a combo-condition using the logical OR operator (i.e. "||") implemented in each of two parallel IFs structures:

- The Endless Loop starts with the usual check of the "Remocon Data Received" flag (Line 16), next it saves the most recent RC-100 input message into Parameter Data (Line 17) and prints it (Line 18).

- Button1 is extracted out of Parameter Data by ANDing it with the Mask U+D+L+R (Line 20 in Fig. 3.25 and more details in Fig. 3.26). In this Figure, "Example Data" shows that Buttons "L" and "3" were pushed, but after the "&" operation with the Mask "U+D+L+R", only the information from Button "L" remains. More details are explained in Video 3.21.

- Similarly, Button2 is extracted out of Parameter Data by ANDing it with the Mask 1+2+3+4 (Line 22 in Fig. 3.25 and more details in Fig. 3.27). In this Figure, "Example Data" shows that Buttons "L" and "3" were pushed, but after the "&" operation with the Mask "1+2+3+4", only the information from Button "3" remains.

- Next, two parallel IFs are used to check on Button1 and Button2 separately for their actual "values", i.e. membership within their appropriate groups/masks (Lines 23 and 28). If each or both IFs turn out to be true, then the values for Button1 or/and Button2 would be printed along their appropriate markers "11" or "22".

```
ENDLESS LOOP
{
    WAIT WHILE ( Remocon Data Received == FALSE (0) )
    Data = Remocon RXD
    Print Screen with Line = Data
    // Filtering out U-D-L-R buttons for Button1
    Button1 = Data & U+D+L+R
    // Filtering out 1-2-3-4 buttons for Button2
    Button2 = Data & 1+2+3+4
    IF ( Button1 == U || Button1 == D || Button1 == L || Button1 == R )
    {
        Print Screen = Button1
        Print Screen with Line = 11
    }
    IF ( Button2 == 1 || Button2 == 2 || Button2 == 3 || Button2 == 4 )
    {
        Print Screen = Button2
        Print Screen with Line = 22
    }
    CALL Delay
}
```

Fig. 3.25 Main algorithm in program “RC-2Buttons.tskx”.

AND Operation with Mask U-D-L-R

Bit Position	*9*	*8*	*7*	*6*	*5*	*4*	*3*	*2*	*1*	*0*
Button	6	5	4	3	2	1	R	L	D	U
Example Data	0	0	0	1	0	0	0	1	0	0
						&				
Mask U+D+L+R	0	0	0	0	0	0	1	1	1	1
Result from & Op.	***0***	***0***	***0***	***0***	***0***	***0***	***0***	***1***	***0***	***0***

A	B	A & B
1	0	0
1	1	1
0	0	0
0	1	0

Fig. 3.26 Result from ANDing operation between "Example Data" and Mask (U+D+L+R).

AND Operation with Mask 1-2-3-4

Bit Position	*9*	*8*	*7*	*6*	*5*	*4*	*3*	*2*	*1*	*0*
Button	6	5	4	3	2	1	R	L	D	U
Example Data	0	0	0	1	0	0	0	1	0	0
						&				
Mask 1+2+3+4	0	0	1	1	1	1	0	0	0	0
Result from & Op.	***0***	***0***	***0***	***1***	***0***	***0***	***0***	***0***	***0***	***0***

A	B	A & B
1	0	0
1	1	1
0	0	0
0	1	0

Fig. 3.27 Result from ANDing operation between "Example Data" and Mask (1+2+3+4).

Video 3.21 shows the run-time performance of this program "RC-2Buttons.tskx".

3.7.2 RC of "Avoider" using PC's Virtual RC-100

We are now ready to expand the concepts learned from Section 3.7.1 to work on a Remote-Control project for the Avoider which allows the user to use multiple Direction Buttons **simultaneously** among the group U-D-L-R and to additionally set Low-Speed or High-Speed parameters on the fly using Buttons 1 and 3 ("RC-MultiDirectionsSpeeds.tskx).

The first part of the main algorithm is shown in Fig. 3.28:

```
20  ENDLESS LOOP
21  {
22      WAIT WHILE ( Remocon Data Received == FALSE (0) )
23      Data = Remocon RXD
24      Print Screen with Line = Data
25      IF ( Data == -- )
26      {
27          CALL Stop
28      }
```

Fig. 3.28 First part of algorithm used in "RC-MultiDirectionsSpeeds.tskx".

- After the usual "Remocon Data Received" flag checking and the reading/printing of the incoming message (Lines 22-24), and if the incoming Data turns out to be "0" (line 25), i.e. "No Button" is pushed, "CALL Stop" is used to stop all motions for the robot (Line 27).

- If Line 25 turns out to be FALSE, i.e. some Buttons are pushed, then the ELSE branch is taken (Line 29) and each possible Direction and Speed Buttons are parsed out using the "&" Operator with the appropriate masks (see Fig. 3.29, Lines 32-37).

```
29      ELSE
30      {
31          // Filtering out specific U-D-L-R-1-3 buttons
32          ButtonUp = Data & U
33          ButtonDown = Data & D
34          ButtonLeft = Data & L
35          ButtonRight = Data & R
36          Button1 = Data & 1
37          Button3 = Data & 3
```

Fig. 3.29 Parsing out various Direction and Speed Buttons in "RC-MultiDirectionsSpeeds.tskx".

• Next, 6 parallel IFs are used to trigger the appropriate maneuvers and speed settings depending on which Buttons are determined to be pushed simultaneously at run-time (see Fig. 3.30).

Video 3.22 shows the runtime performance of the program "RC-MultiDirectionsSpeeds.tskx" where ones can observe that Speed settings can be made in parallel with the directional maneuvers of the robot. While the maneuvers themselves can be combined with each other in a functional way (e.g. "U" and "R" together) or in a completely ineffective way (e.g. "U" and "D" together)!

```
IF ( ButtonUp == U )
{
    CALL Forward
}
IF ( ButtonDown == D )
{
    CALL Backward
}
IF ( ButtonLeft == L )
{
    CALL Left
}
IF ( ButtonRight == R )
{
    CALL Right
}
IF ( Button1 == 1 )
{
    Speed = 400
}
IF ( Button3 == 3 )
{
    Speed = 800
}
```

Fig. 3.30 Parallel IFs used for setting Direction(s) and Speed in "RC-MultiDirectionsSpeeds.tskx".

3.7.3 RC of "Avoider" using Mobile Virtual Controllers

The user may prefer to use the Mobile Virtual Controller instead due to its portability and the better "Human-Machine-Interface" provided by the "in-between" Buttons or by its Joystick or Tilt-Sensor modes (see Figs. 2.9, 2.10 and 2.11). The Mobile Virtual Controller also has "sticky" Buttons 5 and 6 which can be handy for robot games and competitions. The program "RC-MultiDirectionsSpeeds-Mobile.tskx" shows how to program RC maneuvers using those "in-between" Buttons such as "U+R" or "D+L" with the user's left thumb, while letting the user change the speed with Buttons "1" and "3" using the right thumb. The key idea is to treat these "in-between" Buttons as if they were regular "single" Button like the individual U-D-L-R buttons. The main steps of the algorithm used are described below:

- In a similar manner as previously shown in Fig. 3.29, this algorithm needs first to parse out the separate information regarding Direction, Button1 and Button3 (see Lines 29-31 in Fig. 3.31).

```
26  ELSE
27  {
28      // Filtering out specific Direction buttons or 1-3 buttons
29      Direction = Data & [U+D+L+R]
30      Button1 = Data & [1]
31      Button3 = Data & [3]
```

Fig. 3.31 Parsing out Direction and Speed Buttons in "RC-MultiDirectionsSpeeds-Mobile.tskx".

- Next, the previously used "Parallel IFs" structure is replaced by an IF-ELSE-IF structure for the Direction task (see Fig. 3.32), as we are now dealing with a "one and only one alternative" situation, and not with possible simultaneous alternatives like in the previous Section 3.7.2.

Video 3.23 shows the run-time performance of this program "RC-MultiDirectionsSpeeds-Mobile.tskx" off a Tablet.

```
IF ( Direction == U )
{
    CALL Forward
}
ELSE IF ( Direction == D )
{
    CALL Backward
}
ELSE IF ( Direction == L )
{
    CALL Left
}
ELSE IF ( Direction == R )
{
    CALL Right
}
ELSE IF ( Direction == U+R )
{
    CALL Forward
    CALL Delay
    CALL Right
    CALL Delay
}
ELSE IF ( Direction == U+L )
{
    CALL Forward
    CALL Delay
    CALL Left
    CALL Delay
}
```

Fig. 3.32 IF-ELSE-IF used for setting Direction(s) in "RC-MultiDirectionsSpeeds-Mobile.tskx".

3.7.4 Combining Autonomous Behavior and Remote Control

This section shows how to combine Autonomous Behavior and Remote Control techniques so that the Avoider robot can "save itself" when, on purpose, the user drives the robot straight into an obstacle, i.e. we'll try to program a little bit of "intelligence" into the robot!

This program adapts the previous code "RC-MultiDirectionsSpeeds-Mobile.tskx" into "SmartAvoider.tskx" with 4 algorithm changes/additions:

1) A new Variable "Threshold" is used on both IR Left and IR Right Sensors to tell the robot when to start turning itself away from the oncoming obstacle. It is initially set to 100, see Line 11 of Fig. 3.33.

```
6   // Init Action
7   Buzzer Timer  =  Melody Time
8   Buzzer Index  =  Melody No.3 (3)
9   WAIT WHILE ( Buzzer Timer > 0.0sec )
10  Speed = 256
11  Threshold = 100
12  Data = 0
13  DelayTime = 250
14  Direction = 0
15  Button1 = 0
16  Button3 = 0
```

Fig. 3.33 New Variable "Threshold" used to decide when to avoid obstacle in "SmartAvoider.tskx".

2) The first substantial algorithmic change is to process the RC inputs coming in from the user only when **there is no obstacle** registered by both IR Left and Right Sensors (i.e. when the condition (IR Left < Threshold).AND.(IR Right < Threshold) is TRUE). This condition is represented in the IF construct by Line 29 of Fig. 3.34.

3) As before, this algorithm supports the changing of the robot's Speed via Buttons 1 and 3 (see Lines 82 and 87 of Fig. 3.35) but adds the appropriate change in value for Variable Threshold (Lines 83 and 88 of Fig. 3.35), as "physically" the robot needs to start avoiding the obstacle at a farther distance (i.e. a lower valued Threshold), when its Speed is set with a higher value.

```
IF ( Data == -- )
{
    CALL Stop
}
ELSE
{
    IF ( IR Left < Threshold && IR Right < Threshold )
    {
        // No obstacle in front so OK to abide by user's RC inputs
        // Filtering out specific Direction buttons or 1-3 buttons
        Direction = Data & U+D+L+R
        Button1 = Data & 1
        Button3 = Data & 3
        IF ( Direction == U )
        {
            CALL Forward
        }
        ELSE IF ( Direction == D )
        {
            CALL Backward
        }
```

Fig. 3.34 Processing user's RC Inputs **only when** no obstacle located in front of robot ("SmartAvoider.tskx").

```
IF ( Button1 == 1 )
{
    Speed = 400
    Threshold = 70
}
IF ( Button3 == 3 )
{
    Speed = 800
    Threshold = 30
}
```

Fig. 3.35 Appropriate value setting of "Threshold" depending on "Speed" ("SmartAvoider.tskx").

4) When the condition **(IR Left < Threshold).AND.(IR Right < Threshold)** (Line 29 in Fig. 3.34) evaluates to **FALSE**, this means that there is an obstacle in front of either or both IR Sensors, then the algorithm goes to its corresponding ELSE (Line 94 of Fig. 3.36) where a new section of code is created to handle the "automatic" obstacle avoidance maneuvers:

```
94  ELSE
95      // Obstacle is getting close - Robot stops first then tries to avoid obstacle on its own
96  {
97      CALL Stop
98      LOOP WHILE ( IR Left > 10 || IR Right > 10 )
99      {
100         LOOP WHILE ( IR Left > 5 )
101         {
102             CALL Right
103         }
104         LOOP WHILE ( IR Right > 5 )
105         {
106             CALL Left
107         }
108     }
109     // Here, no more obstacle in front - Robot stops and wait for user to release buttons on RC-100
110     CALL Stop
111     LOOP WHILE ( Data != -- )
112     {
113         WAIT WHILE ( Remocon Data Received == FALSE (0) )
114         Data = Remocon RXD
115     }
116 }
```

Fig. 3.36 Automatic Obstacle Avoidance Procedure in "SmartAvoider.tskx".

- First the robot is stopped by "CALL Stop" (Line 97 of Fig. 3.36).
- The outer WHILE Loop (Lines 98 to 108) is designed so that as long **"IR Left" or "IR Right"** indicate some sort of obstacle, i.e. when Condition ((IR Left > 10).OR.(IR Right > 10)) is TRUE, the robot should keep on APPROPRIATELY turning RIGHT (Lines 100-103) or LEFT (Lines 104-107) depending on which IR Sensor reads a value more than "5". This algorithm does favor "turning right" when the robot runs straight on into an obstacle, which is a more likely maneuver if the robot "resides" in countries where cars are driven on the right side of the road!
- When the algorithm gets out of the outer WHILE Loop at Line 109, "back in the physical world" everything is clear in front of the robot, then the robot stops and just waits there for the user to "gather his/her wits" after such a close call! At that time, assuming the user was still pressing on

some button(s) on the Remote Controller, the last WHILE loop (Lines 111-115) is designed to wait the user to release all buttons on the RC-100, i.e. it keeps on looping as long as (Data NOT EQUAL to "No Button") – conditional statement used on Line 111 (and the author apologizes for using a "double negative" here, but it was necessary!). An alternate solution is offered in the Review Questions Section of this Chapter.

- When the user finally releases all RC buttons, the algorithm goes back to the original ENDLESS LOOP, and essentially allows the user to use the RC-100 once again to move the robot around.

Video 3.24 shows the run-time performance of the program "SmartAvoider.tskx".

3.8 Line Tracking & Obstacle Avoidance with NIR Sensors

For this Section 3.8, we will be using a CarBot configuration as described in Fig. 3.37. Construction wise, this CarBot derives from the Avoider (Fig. 3.10), but the whole robot gets rotated down 90 degrees so that the Left & Right IR Sensors are now aimed towards the ground and with additional "sprocket" wheels for front end support. The goal is to program this CarBot to go forward and follow an arbitrary track created using standard black electrical tape laid out on brown Craft paper. This CarBot would have to use its Left and Right NIR sensors to keep it "on-track". During its progress along this track, if the CarBot finds a frontal obstacle, it should be able to swing around, re-discover the black track and follow it again but in the opposite direction.

Fig. 3.37 CarBot used as Autonomous Line Tracker & Obstacle Avoider.

Let's first analyze how the three built-in NIR Sensors behave when the CarBot is laid on the black track under the four main conditions that would occur in this challenge: a) when the track is under the bot's Left IR Sensor; b) when the bot is centered on the track; c) when the track is under the bot's Right IR Sensor; and d) when the bot is gone completely off-track (see Fig. 3.38).

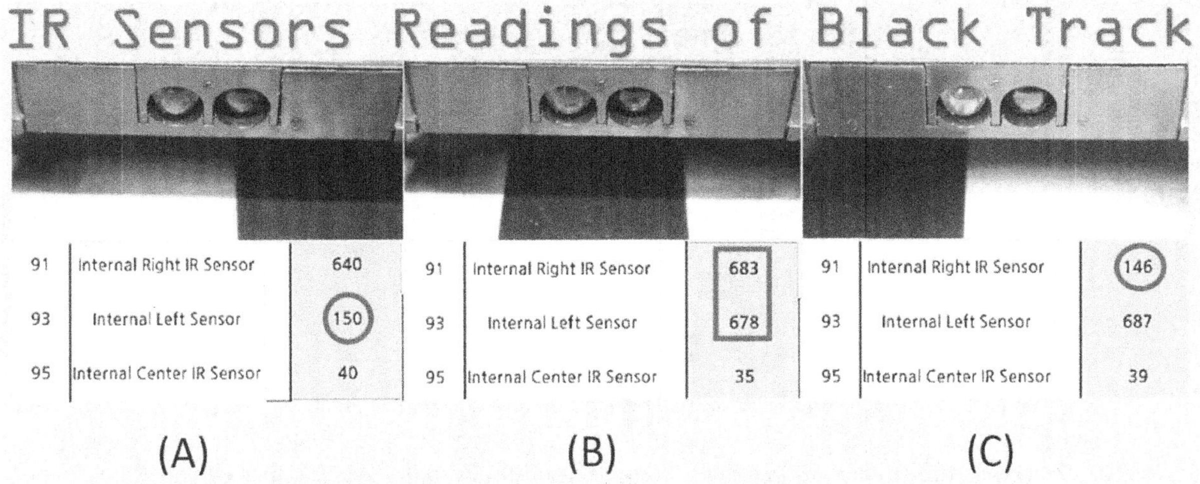

Fig. 3.38 Values for L-R-C IR Sensors for 3 configurations between CarBot and Black Track.

In Fig. 3.38 and going from Column (A) to Column (C) to check for the readings from the Right IR Sensor (Address 91): ones can see that its value is around 640-683 whenever it sees the background Craft paper, and that its value is reduced to around 146 when this sensor is "seeing" the black track. Similarly, for the Left IR Sensor (Address 93), it reads around 150 when it "sees" the black track and between 678 and 687 when it "sees" the Craft paper surface. Most importantly, Column (B) shows that the spacing between the Left and Right IR Sensors is larger than the width of the black electrical tape as both sensors read "high" here (e.g. 683 and 678) - in other words, both sensors are only seeing the brown paper in this situation. **Thus, it is physically impossible for this CarBot to figure out via its Left and Right NIR sensors whether it is centered on the black track (case b) or that it is completely off-track (case d) – not shown in Fig. 3.38!**

Also, in Fig. 3.38, an analysis of the IR Center Sensor data shows that this CarBot was aiming a bit low as it still can see a bit of the "brown paper" resulting in values between 35 and 40.

The previous experimental analysis shows that there are only two conditions (mutually exclusive) that can be monitored unequivocally by this CarBot hardware configuration:

1) **If the Left IR Sensor can see the black track then it should turn left in its attempt to re-align the bot with the black track.**

2) **If the Right IR Sensor can see the black track then it should turn right in its attempt to re-align the bot with the black track.**

Otherwise, this robot really cannot be "sure" of anything else, and so it should continue its default maneuver (i.e. going forward). From previous sections, we are then expecting to see an IF-ELSE-IF structure for three mutually exclusive alternatives in the programming solution for this challenge.

Fig. 3.39 shows the Initial Actions for the solution program "LineFollower-ObstacleAvoider.tskx", whereas the CM-150 plays Melody 3 (Lines 5-7) then sets initial values for 4 Parameters - Speed,

TimeDelay, LineDistance and ObjectDistance (Lines 8-11). The "200" value assigned to Parameter LineDistance is just a number between the values "680" and "140" that were reported for both IR Left and Right Sensors (see Fig. 3.38 and blue circles in columns (A) and (C)). "ObjectDistance" is set to "100", a number sufficiently larger than the typical 35-40 values found in Fig. 3.38.

```
4   // Init Action
5   Buzzer Timer = Melody Time
6   Buzzer Index = Melody No.3 (3)
7   WAIT WHILE ( Buzzer Timer > 0.0sec )
8   Speed = 350
9   TimeDelay = 125
10  LineDistance = 200
11  ObjectDistance = 100
```

Fig. 3.39 Initial Actions for program "LineFollower-ObstacleAvoider.tskx".

Fig. 3.40 shows the EndLess Loop where the main algorithm is found:

1) If a frontal object is found (Line 16), the CarBot should stop then swing around to the reverse direction (Lines 18-19).

2) Otherwise, the programming logic flows into an IF-ELSE-IF structure whereas:

- If only the Left IR Sensor "sees" black (Line 24) then the "LeftTurn" Function is called (Line 26). Otherwise, the Right IR Sensor is checked next (Line 29), and if only this Right Sensor "sees" black then Function "RightTurn" is called (Line 32).
- For all other situations (i.e. the ELSE on line 34), Function "Forward" is executed (Line 37).

```
ENDLESS LOOP
{
    //  If there is an obstacle in front (using IR Center)
    IF  (  IR Center  >  ObjectDistance  )
    {
        CALL  Stop
        CALL  SwingAround
    }
    ELSE
        //  No obstacle in front of CarBot, so OK to follow black track
    {
        IF  (  IR Left  <  LineDistance  &&  IR Right  >=  LineDistance  )
        {
            //  Turns left when ONLY the left sensor detects black.
            CALL  LeftTurn
        }
        ELSE IF  (  IR Left  >=  LineDistance  &&  IR Right  <  LineDistance  )
        {
            //  Turns right when ONLY the right sensor detects black
            CALL  RightTurn
        }
        ELSE
        {
            //  Moves forward for all other cases.
            CALL  Forward
        }
    }
}
```

Fig. 3.40 Main Algorithm for program "LineFollower-ObstacleAvoider.tskx".

The Function "SwingAround" (Fig. 3.41) describes a combination of "timed" and "conditional" right turns to accommodate for the fact that the "obstacle" can be found at any spot on the black track:

1) First, a 0.125s timed Right-Turn is used (Lines 82-85) because at this point in time, the CarBot is mostly lined up with the black track and it just needs to be nudged a bit to the right to make sure that Right IR Sensor is now "seeing" the Craft paper (i.e. "white" - the condition (IR_Right > LineDistance) is TRUE).

```
79  FUNCTION SwingAround
80  {
81      // Turn right for 0.125 second to clear IR-Right sensor off the track
82      PORT[2]:Geared Motor = CCW:0 (0.00%) + 250
83      PORT[1]:Geared Motor = CCW:0 (0.00%) + 250
84      High-resolution Timer = 0.125sec
85      WAIT WHILE ( High-resolution Timer > 0.000sec )
86
87      // Keep on turning right as long as IR-Right sees white
88      WAIT WHILE ( IR Right > LineDistance )
89
90      // IR-Right now sees Black, but OK to keep on turning right until IR-Right sees White again
91      WAIT WHILE ( IR Right <= LineDistance )
92      // Bot should be lining up with the track but in the reverse direction at this point
93  }
```

Fig. 3.41 Function "SwingAround" for program "LineFollower_ObstacleAvoider.tskx".

2) Next, the reader needs to recall that the previous commands for the CarBot to turn right (Lines 82-83) are still in effect. The 2 WHILE loops (Lines 88 and 91) are "conditional" loops so that the CarBot can adjust to "any" section of the arbitrary track where the "SwingAround" maneuver may be needed. These 2 loops are used to ensure that an orderly process is followed to detect whether the Right IR Sensor is seeing "black" or "white" (e.g. brown Craft paper):

- The first loop (Line 88) is active whenever Right IR sees "white", i.e. whenever the CarBot has **not** yet crossed back into the black track. When this situation happens, i.e. IR Right sees "black", the first loop's conditional expression becomes false and this loop is exited, and the second loop is entered (Line 91).

- The purpose for the second loop (Line 91) is to keep the CarBot turning right some more until IR Right sees "white" (i.e. it has just passed over the other edge of the black track). At this point in time, the CarBot should be mostly lined up with the black track (but in the reverse direction), thus this Function is exited and another iteration of the Main Endless Loop is executed, resulting in the Carbot's default maneuver of Going Forward.

Video 3.25 shows the performance of this algorithm on different shapes of a closed circuit black track. Ones can see that this algorithm cannot handle "sharp" turns, as it would get off-track at these spots.

3.9 Using Servo Motor & External NIR Sensor

So far on the CM-150, we have only used the Ports 1 and 2 to interface with the two Geared Motors. The CM-150 also supports Ports 3 and 4 which are more general-purpose ports as they can interface with sensors such as NIR, Color or Passive IR (http://support.robotis.com/en/product/auxdevice/sensor_main.htm), and also with Servo Motors (http://support.robotis.com/en/product/auxdevice/motor/servomotor_manual.htm).

3.9.1 Tricycle Steering Control with Servo Motor

A Servo Motor's operation is very different from the one for the Geared Motor that we had been using. Essentially, a Geared Motor can turn Clockwise or Counterclockwise and at different speeds, while a Servo Motor is used if the user wants to hold a Servo Motor (and the device attached to it) in a programmable but fixed position. Fig. 3.42 looks at the front of a Servo Motor that has an NIR Sensor attached to it. The Servo Motor's position can be controlled by sending it a positive integer between 0 and 1023 whereas the Position Number "512" corresponds the Default "0 degree" Position, while "819" corresponds to +90 degrees and "205" corresponds to -90 degrees. Please watch Video 3.26 for basic use of a Servo Motor.

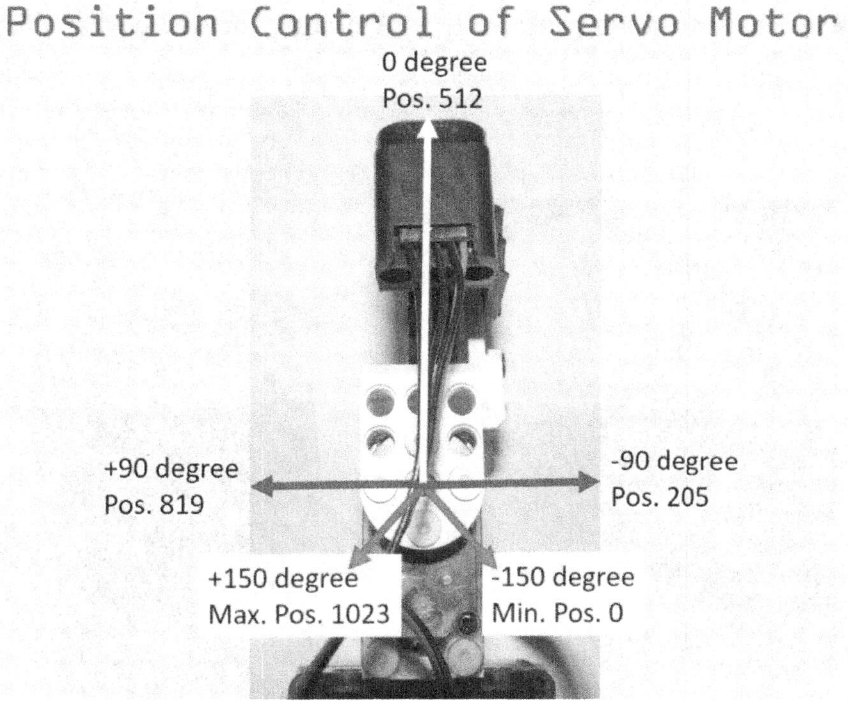

Fig. 3.42 Position Control of Servo Motor with attached NIR Sensor.

The projects in this section use a Servo Motor via Port 3 and an external NIR Sensor via Port 4. The robot platform used is the TriCycle as shown in Fig. 3.43. The TriCycle's powered back end is the same as the previous CarBot (Fig. 3.37) minus the sprocket wheels, but on its front end is mounted a Servo Motor/Front Wheel. The Front Wheel is passive and it is attached via a U-shaped frame to the horn of the Servo Motor. The NIR Sensor is then attached to the Front Wheel assembly (see Fig. 3.44). Thus, the

Servo Motor (Port 3) controls the direction of the Steering Wheel and the View Angle of the NIR Sensor (itself connected to Port 4).

Fig. 3.43 TriCycle robot with passive Steering Wheel and forward-looking NIR Sensor.

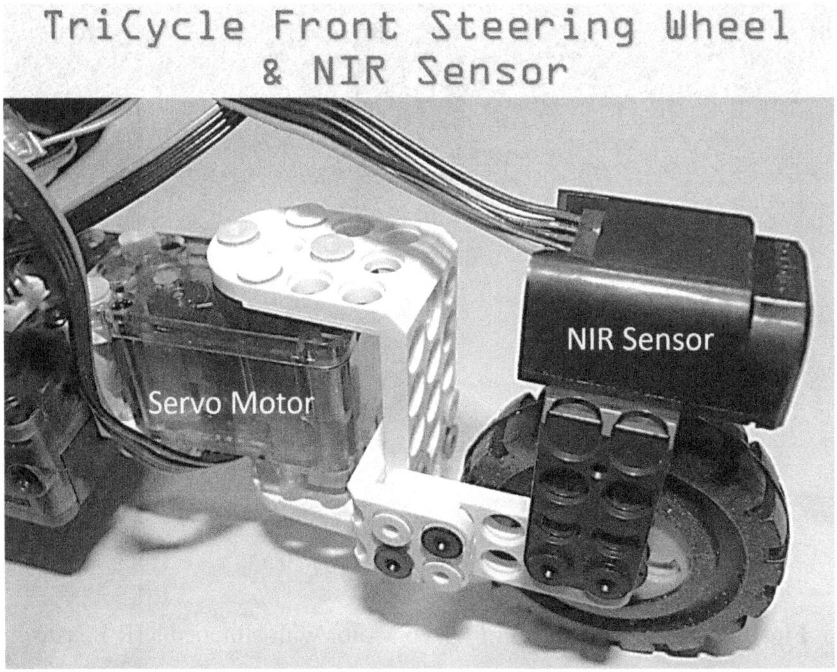

Fig. 3.44 Details of TriCycle's Front End.

Although the SmartAvoider (Section 3.7.4) and the TriCycle share the same triangular shape for their ground contact points (see Fig. 3.45), the SmartAvoider has a more compact triangular shape allowing it to be nimble during turn maneuvers. The TriCycle has an elongated triangular shape, i.e. a longer moment arm between the rear drive wheels and front steering wheel (so-called "Delta" configuration). Furthermore, its rubber front wheel can generate more sideways friction forces with the ground surface than the little plastic support used by the SmartAvoider, so mechanically the TriCycle won't be as nimble as the SmartAvoider. Thus, we will need a different way of activating the Left and Right drive motors, more gradually in conjunction with the front wheel's turn angle, to assist the TriCycle during its Left and Right turns.

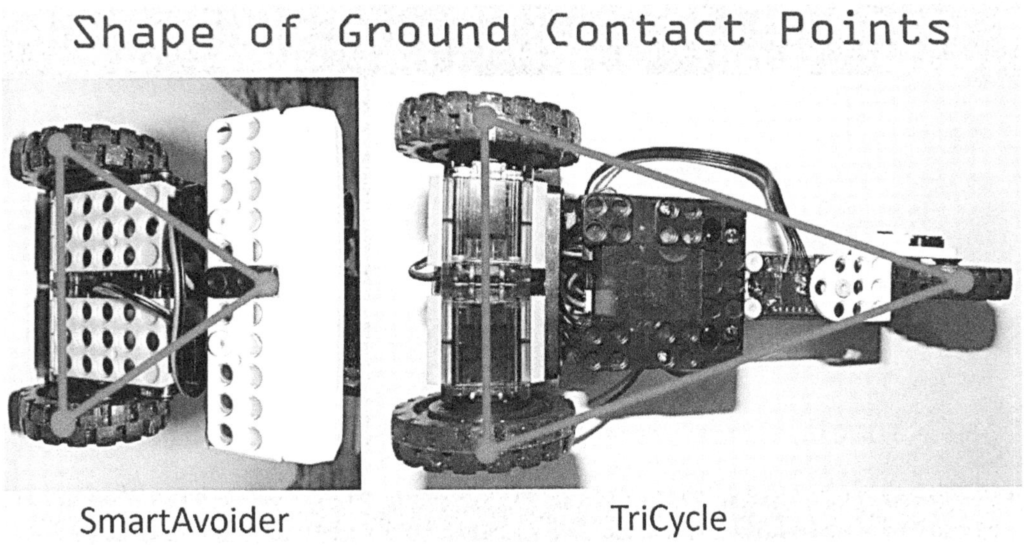

Fig. 3.45 Triangular Shapes of Ground Contact Points for SmartAvoider and TriCycle.

The first TriCycle project is about its Remote Control operations but with a new control approach as implemented in the program "RC-TriCycle.tskx", which itself is based on the previous program "RC-MultiDirectionsSpeeds-Mobile.rskx":

1) The INIT section (see Fig. 3.46) shows how to set up and use the Servo Motor out of Port 3 programming. Line 10 sets the Servo's Drive Mode to TRUE (1), i.e. puts it in Control Position mode *(Note: set this mode to FALSE (0) to use this Servo Motor like a regular Geared Motor)*. Next, Line 11 sets the Servo's Speed to 1023 out of an allowable range of [0-1023]. This Speed controls how fast the Servo can move between programmed Positions which are also in the same [0-1023] range (see their corresponding physical positions as illustrated in Fig 3.42). This Speed value also is a measure of how strong a torque this servo can hold at a programmed position, under the weight/inertia of whatever structure/device that is attached to the same servo. The user could use a mid-value for Speed to achieve less sudden Servo Positions and thus to reduce structure vibrations, but the user needs to use Speed=1023 to hold a steady Servo Position once the servo gets where it is supposed to be. Once the Mode and Speed parameters are set (Lines 10-11), the user can put the Servo Motor to good use by setting its next Control Position value (Line 12) which should be an integer between 0 and 1023 (see Fig. 3.42). There is one important point that beginner users need to be cognizant of when programming servo positional changes: as it will take some actual time periods for the servo to get from one physical location to the next, the user needs to program in some WAIT time for the current operation to finish, before issuing a new command to move the same servo to another location.

2) The INIT section also contains 3 new Variables: SpeedL, SpeedR and SpeedDif (see Fig. 3.46). SpeedL is used exclusively by the Left Geared Motor (Port 1), while SpeedR is reserved for the Right Geared Motor (Port 2). Please see Fig. 3.47 for the use of SpeedL and SpeedR in the new Functions

Forward and Backward as examples. SpeedDif will be set proportionally to the Turn Angle of the Steering Wheel's Servo Motor (controlled via Port 3) when Left or Right Turns are required.

```
PORT[3]:Servo Motor Drive Mode = TRUE (1)
PORT[3]:Servo Motor Speed = 1023
PORT[3]:Servo Motor Position = 512
Speed = 256
SpeedL = 0
SpeedR = 0
SpeedDif = 0
Data = 0
```

Fig. 3.46 Init Event with three new Variables: SpeedL, SpeedR & SpeedDif in "RC-TriCycle.tskx"

```
FUNCTION Forward
{
    PORT[2]:Geared Motor = CW:0 (0.00%) + SpeedR
    PORT[1]:Geared Motor = CCW:0 (0.00%) + SpeedL
    RETURN
}

FUNCTION Backward
{
    PORT[2]:Geared Motor = CCW:0 (0.00%) + SpeedR
    PORT[1]:Geared Motor = CW:0 (0.00%) + SpeedL
    RETURN
}
```

Fig. 3.47 Examples of usage of SpeedL and SpeedR in "RC-TriCycle.tskx"

3) Next the procedure to parse the Remocon Packet "Data" into separate Parameters "Direction", "Button1" and "Button3" remains the same, but the Actions that correspond to the cases when (Direction == "U") or when (Direction == "D") are quite different (see Fig. 3.48). Let's take the case when (Direction == "U") is TRUE in Line 39, then Port 3 Servo Position is set to "512" (Line 41- i.e. Front Steering Wheel set to go straight) and SpeedDif is set to "0" (Line 42), thus SpeedL and SpeedR would have the same numerical value as Parameter Speed (which can be either "400" or "800") – see Lines 43 and 44. Finally Function Forward is called to make the TriCycle go Forward (Line 45 of Fig.3.48). A similar sequence of actions would be executed when (Direction == "D") is TRUE in Line

47, whereas the TriCycle would go Backward in a "straight" line (Lines 49 to 53 in Fig. 3.48). If for some reasons, there are some mechanical inaccuracies in the robot assembly, "SpeedDif" may have to be a non-zero value and need to be manually adjusted to make this TriCycle go really "straight" in its forward or backward motions (the user would need to experiment with his/her own TriCycle robot to obtain the optimal value for SpeedDif).

```
39  IF ( Direction == U )
40  {
41      PORT[3]:Servo Motor Position = 512
42      SpeedDif = 0
43      SpeedL = Speed - SpeedDif
44      SpeedR = Speed + SpeedDif
45      CALL Forward
46  }
47  ELSE IF ( Direction == D )
48  {
49      PORT[3]:Servo Motor Position = 512
50      SpeedDif = 0
51      SpeedL = Speed - SpeedDif
52      SpeedR = Speed + SpeedDif
53      CALL Backward
54  }
```

Fig. 3.48 Sequence of Actions executed when "Direction" == "U" or "D" in "RC-TriCycle.tskx"

4) When (Direction == "L" or "R"), i.e. when Left or Right Turns are required, the appropriate sequences of required Actions are quite a bit different than for Forward or Backward (see Fig. 3.49). For example, Line 57 shows that **each time** that (Direction == "L") is TRUE, the **Current Servo Position is increased by "50" units** which make the **Steering Wheel turn Left about an "extra" 15-degrees angle each time**. In other words, if the user keeps on pressing the "L" button when wanting to make a Left Turn, like in all previous applications, this TriCycle will make tighter and tighter turns "on its own", so the user needs to adjust his or her "driving" behavior for this "new design". Furthermore, "SpeedDif" is also increased by "5" units each time that the condition (Direction == "L") is TRUE (see Line 58). Next, Lines 59 and 60 shows that SpeedL is decreased by SpeedDif (i.e. Left Motor slows down) while SpeedR is increased by SpeedDif (i.e. Right Motor speeds up), i.e. making the robot turn Left. In other words, the Left and Right Motors are also differentially commanded to assist in the Left Turn motion initiated by the Steering Wheel, because the TriCycle has a longer moment arm to work with as compared to the Avoider that had been used so far in this book (see Fig. 3.45).

```
55  ELSE IF ( Direction == L )
56  {
57      PORT[3]:Servo Motor Position = PORT[3]:Servo Motor Position + 50
58      SpeedDif = SpeedDif + 5
59      SpeedL = Speed - SpeedDif
60      SpeedR = Speed + SpeedDif
61      CALL CheckSpeeds
62      CALL Left
63  }
64  ELSE IF ( Direction == R )
65  {
66      PORT[3]:Servo Motor Position = PORT[3]:Servo Motor Position - 50
67      SpeedDif = SpeedDif + 5
68      SpeedL = Speed + SpeedDif
69      SpeedR = Speed - SpeedDif
70      CALL CheckSpeeds
71      CALL Right
72  }
```

Fig. 3.49 Sequence of Actions executed when "Direction" == "L" or "R" in "RC-TriCycle.tskx"

As the substracting and adding operations in Lines 59-60 may make the Parameters SpeedL and SpeedR go out of their "allowed" range which is between 0 and 1023, the Function CheckSpeeds (Line 61 of Fig. 3.49) is always executed (see code details in Fig. 3.50) before the Function Left is called to perform the actual Left Turn maneuver in Line 62 of Fig 3.49.

A similar procedure is used when the condition (Direction == "R") is TRUE to make the TriCycle performs an increasing Right Turn. The reader can watch Video 3.27 to see the TriCycle do a "pop-wheelie" during these maneuvers.

```
FUNCTION CheckSpeeds
{
    IF ( SpeedL < 0 )
    {
        SpeedL = 0
    }
    ELSE IF ( SpeedL > 1023 )
    {
        SpeedL = 1023
    }
    IF ( SpeedR < 0 )
    {
        SpeedR = 0
    }
    ELSE IF ( SpeedR > 1023 )
    {
        SpeedR = 1023
    }
}
```

Fig. 3.50 Function CheckSpeeds to clamp down SpeedL and SpeedR to the appropriate limits if they get out of range [0-1023] in "RC-TriCycle.tskx".

5) Last but not least, the maneuvers for the "in-between" positions like U+R, U+L, D+R and D+L are also implemented very differently for the TriCycle from what they were in previous RC applications (see Figs. 3.51 and 3.52).

For example, in Fig. 3.51 (Lines 75-78), when "U+R" is TRUE, the current Servo Position is increased by 25 only as compared to the previous cases for "R" and "L" (i.e. a "milder" change in Steering Wheel angle is used). SpeedDif also increases only by 3 (i.e. a "milder" change in Motor Speed is also used in conjunction with "milder" Steering changes). "CheckSpeeds" is still used to clamp down SpeedL and SpeedR to the "legal" limits if needed (Line 79). Please note that only "CALL Forward" is used with a 0.25 second delay to create "smooth" and "mild" maneuvers.

A similar procedure is used for "U+L" in Lines 83-92 of Fig. 3.51 to create "smooth and mild" Left turns.

Similar maneuvers are also implemented for the other cases "D+R" and "D+L", except now the TriCycle will use the Function Backward instead (see Fig. 3.52).

```
ELSE IF ( Direction == U+R )
{
    PORT[3]:Servo Motor Position = PORT[3]:Servo Motor Position - 25
    SpeedDif = SpeedDif + 3
    SpeedL = Speed + SpeedDif
    SpeedR = Speed - SpeedDif
    CALL CheckSpeeds
    CALL Forward
    CALL Delay
}
ELSE IF ( Direction == U+L )
{
    PORT[3]:Servo Motor Position = PORT[3]:Servo Motor Position + 25
    SpeedDif = SpeedDif + 3
    SpeedL = Speed - SpeedDif
    SpeedR = Speed + SpeedDif
    CALL CheckSpeeds
    CALL Forward
    CALL Delay
}
```

Fig. 3.51 Maneuvers programmed for "U+R" and "U+L" inputs from the user in "RC-TriCycle.tskx".

```
ELSE IF ( Direction == D+R )
{
    PORT[3]:Servo Motor Position = PORT[3]:Servo Motor Position - 25
    SpeedDif = SpeedDif + 3
    SpeedL = Speed + SpeedDif
    SpeedR = Speed - SpeedDif
    CALL CheckSpeeds
    CALL Backward
    CALL Delay
}
ELSE IF ( Direction == D+L )
{
    PORT[3]:Servo Motor Position = PORT[3]:Servo Motor Position + 25
    SpeedDif = SpeedDif + 3
    SpeedL = Speed - SpeedDif
    SpeedR = Speed + SpeedDif
    CALL CheckSpeeds
    CALL Backward
    CALL Delay
}
```

Fig. 3.52 Maneuvers programmed for "D+R" and "D+L" inputs from the user in "RC-TriCycle.tskx".

The user may find that a user interface like the Joystick on the mobile R+m.TASK would be a better "Human-Machine-Interface" for this TriCycle application as the JoyStick allows a more "continuous" finger motion within its ring-shaped touch area. Please watch Video 3.27 for a demonstration of this TriCycle.

Among the accompanying example codes, the reader can look for and try out another Remote-Control approach whereas the L and R buttons are to be "pulsed" (with audio feedback) for the Left and Right turns to work properly (see "RC-TriCycle-LR_Pulse.tskx"). Furthermore, in this scheme, the buttons U-D, L-R and 1-3 are treated as independent button groups.

The reader is also encouraged to try his or her own control scheme whereas each turn corresponds to a "fixed" angular change for the Steering Wheel and a "fixed" numerical change in the parameter Speed Differential. In the Review Questions Section of this Chapter, the reader is also encouraged to create a Remote-Control scheme for a "Front-Wheel-Drive" TriCycle, i.e. the Directional Wheel (controlled by the Servo Motor) is now considered to be in the rear of the vehicle (i.e. so-called "Tadpole" configuration).

3.9.2 Dowel Scanner

For this project, we would need a nimble robot, so it's back to the Avoider's platform, but this time with a Servo Motor "Head" that is slewing a forward-looking NIR Sensor (see Fig. 3.53), yielding a "Dowel

Scanner" robot. This robot's tasks are to identify two wooden dowels placed at random in front of it (see Fig. 3.54), specifically to identify which one is closest to it and approach it within a certain distance, then it would need to switch its attention to find the second dowel and likewise to approach it within a given distance.

Fig. 3.53 "Dowel Scanner" Robot.

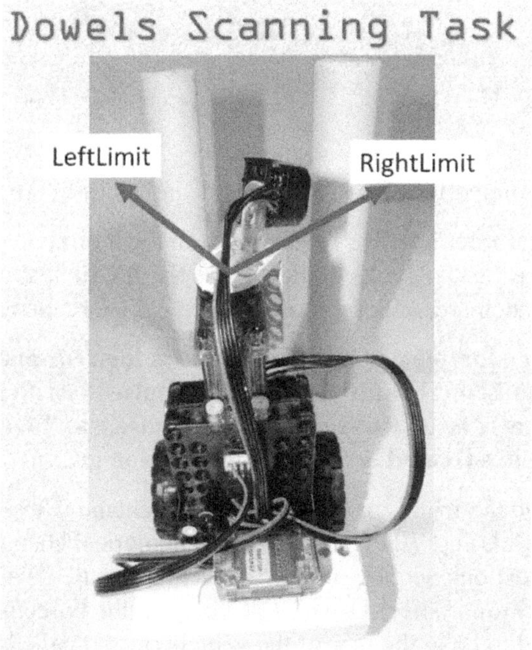

Fig. 3.54 Dowel Scanning Task Layout.

In this project, the NIR Sensor can be slewed in two ways:

1) By the Servo Motor via Port 3 (while the robot is standing still): this technique will be used to identify which dowel (Left or Right) is closest to the robot at the start of the tasks.

2) With the Servo Motor locked into Position 512, thus with the NIR Sensor facing straight forward, the whole robot platform is rotated left-right as needed by the two Geared Motors via Ports 1 and 2 to perform the dowel scanning task using the NIR Sensor, this mode is used in the final identification of each dowel and in the final approach to each dowel.

Figs. 3.53 and 3.54 also show that the NIR Sensor is mounted on its "side" to take full advantage of the fact that the dowels are "vertical" objects. Fig. 3.55 is an output list of the Servo's Position data (first column) and the NIR Sensor's values (second column) at the same positions, as the NIR Sensor is slewed from Left to Right (i.e. Position values from ~700 to ~380), using a simple TASK program.

Program Output Monitor

703	3
705	6
705	3
697	3
685	5
674	4
662	5
650	5
635	62
621	154
609	294
595	289
581	125
567	30
552	6
539	4
523	3
508	5
494	4
479	27
463	37
448	53
433	28
419	7
406	4
392	5
379	3

Fig. 3.55 Data Output for Servo's Positions and corresponding NIR Values as the NIR Sensor is slewed across two wood dowels from Left to Right.

The reader can readily see that the "background" reads as "3 to 7" and that the dowel's presence is easily determined by an increase followed by a decrease in NIR values. The first dowel can be found around Position 609 (i.e. on Left of robot) corresponding to a local maximum NIR value of 294, while the second dowel can be found around Position 448 (i.e. on Right of robot) corresponding to a lower local maximum of 53. Thus, we can also determine that the "Left" dowel (max NIR value = 294) is closer to the robot than the one on its "Right" (max NIR value = 53). This is the strategy used in developing a search algorithm in TASK code to find the two dowels, to be described in later parts of this Section 3.9.2.

The complete program for this project is named "DowelScanner.tskx" and its main parts are described below:

1) Part 1 of The INIT event is described in Fig. 3.56, whereas many variables are initialized. First "Re#" is played for 1 second (Lines 7-9), then Port 3 Servo Motor is defined for Position Control mode (Line 10), with its Speed set to 512 (50% power – Line 11), and its initial Position set to 512 (Line 12). "SpeedL" and "SpeedR" are used for Ports 1 and 2 Geared Motors as before. The LeftLimit and the RightLimit parameters are used to define the slewing range for the Port 3 Servo Motor. The other parameters ProximityLimit, PreviousNIR, CurrentNIR, DowelDistance1, DowelDistance2, Position1, Position2 and FoundDowel will be described and their usage explained in later paragraphs.

```
7   Buzzer Timer = 1.0sec
8   Buzzer Index = Re# (6)
9   WAIT WHILE ( Buzzer Timer > 0.0sec )
10  PORT[3]:Servo Motor Drive Mode = TRUE (1)
11  PORT[3]:Servo Motor Speed = 512
12  PORT[3]:Servo Motor Position = 512
13  SpeedL = 200
14  SpeedR = 200
15  LeftLimit = 720
16  RightLimit = 350
17  ProximityLimit = 0
18  PreviousNIR = 0
19  CurrentNIR = 0
20  DowelDistance1 = 0
21  DowelDistance2 = 0
22  Position1 = 0
23  Position2 = 0
24  FoundDowel = 0
```

Fig. 3.56 Part 1 of the INIT Event in "DowelScanner.tskx".

2) Next, the robot waits for the user to make some noise (clapping hands or loud whistling) (Line 25 of Fig. 3.57). Once the Sound Counter becomes non-zero, "Background" is initialized to zero (Line 26). Next, a FOR loop takes in 20 consecutive Data Values from Port 4 NIR Sensor (Lines 27-30) and adds them up into "BackgroundNIR", next this cumulative sum is averaged out and saved back into the same Parameter "BackgroundNIR" (Line 31). "DowelThreshold" is defined as having 3 units higher than "BackgroundNIR". The final Init Action is to play Melody 0 (Lines 33-35).

```
25  WAIT WHILE ( Result of Sound Counter == 0 )
26  BackgroundNIR = 0
27  LOOP FOR ( I = 1 ~ 20 )
28  {
29      BackgroundNIR = BackgroundNIR + PORT[4]:IR Sensor
30  }
31  BackgroundNIR = BackgroundNIR / 20
32  DowelThreshold = BackgroundNIR + 3
33  Buzzer Timer = Melody Time
34  Buzzer Index = Melody No.0 (0)
35  WAIT WHILE ( Buzzer Timer > 0.0sec )
36
37  CALL ServoScan
38  CALL VerifyDowels
```

Fig. 3.57 Part 2 of the INIT Event and Main Actions in "DowelScanner.tskx".

The Main Actions are to CALL "ServoScan" and "VerifyDowels" (Lines 37-38). "ServoScan" does the first scan using the Port 3 Servo and NIR Sensor combo to figure out which dowel is closest to the robot. "VerifyDowels" accomplishes the other tasks of final identification and robot approach using Ports 1 and 2 Geared Motors in conjunction with the NIR Sensor.

3) Next, let's dive into Function "ServoScan" with Fig. 3.58. First, Port 3 Servo is sent to its "LeftLimit" position (i.e. 720). Next, "LeftLimit" is decreased by 20 units (i.e. to 700) by Line 79 and this new value is used in a WAIT WHILE Loop on Line 80 (essentially making the CM-150 "do nothing" while waiting for the Servo Motor to reach its assigned location of 720). Please note that this first Servo Move is done at Speed = 512 (see Line 11 of Fig 3.56). Once the Servo passes over Position 700, "FoundDowel" is initialized to 0 (Line 81), the Servo Speed is slowed down to 200 (Line 82), and the Servo is once again sent away to Position "RightLimit" (i.e. 350 in Line 16 of Fig. 3.56). This "trip" would take a longer time period to finish and it is during this "trip" that the two dowels would be scanned for by the NIR Sensor and their closeness to the robot determined. This task is performed by Function "FindDowelA" on Line 86 for the first dowel (i.e. the Left one), and its (max) NIR Distance to the robot is saved into Parameter "DowelDistance1". The WAIT WHILE in Line 90 is used for the controller to pass some time, while the NIR Sensor scans over the remainder of the Left Dowel and starts getting into the empty space right after the Left dowel. In this transitional zone, the NIR Sensor would read in values that are greater than "BackgroundNIR". Once the NIR Sensor "sees" the empty space between the two dowels, the controller would quit this WAIT WHILE Loop, and execute Line 93 next to find the second dowel on the Right and save its (max) NIR distance from the robot into Parameter "DowelDistance2". Lines 95 and 96 stop the Servo Motor and "freeze" its Current Position (review Video 3.26). Lines 97 and 98 display the Dowel Distances found for the user to help keeping track of the robot's progress.

```
FUNCTION ServoScan
{
    PORT[3]:Servo Motor Position = LeftLimit
    LeftLimit = LeftLimit - 20
    WAIT WHILE ( PORT[3]:Servo Motor Position < LeftLimit )
    FoundDowel = 0
    PORT[3]:Servo Motor Speed = 200
    PORT[3]:Servo Motor Position = RightLimit

    // Finding Dowel 1
    CALL FindDowelA
    DowelDistance1 = PreviousNIR

    // Let IR Sensor scan by the rest of Dowel 1 and until IR Sensor sees Background again
    WAIT WHILE ( PORT[4]:IR Sensor > BackgroundNIR )

    // Finding Dowel 2
    CALL FindDowelA
    DowelDistance2 = PreviousNIR
    PORT[3]:Servo Motor Speed = 0
    PORT[3]:Servo Motor Position = PORT[3]:Servo Motor Position
    Print Screen = DowelDistance1
    Print Screen with Line = DowelDistance2
}
```

Fig. 3.58 Function "ServoScan" in "DowelScanner.tskx".

4) Function "FindDowelA" (see Fig. 3.59) implements the search strategy briefly mentioned in Paragraph 3). The reader would need to remember that the NIR Sensor started to scan from Servo Position "LeftLimit" just before Function "FindDowelA" is called. This means that the NIR Sensor would read in values mostly around the current value of Parameter "BackgroundNIR" during this time, and as "DowelThreshold" was set to be "3" units above "BackgroundNIR", it was used in the WAIT WHILE on Line 103 to let the controller "do nothing" until the NIR Sensor starts to detect the Left Edge of the Left Dowel. At that time, the NIR Sensor would read in values greater than "DowelThreshold", thus the controller would quit this WAIT WHILE loop, and start on the actual search algorithm alluded to earlier with Line 104. As shown back in Fig. 3.52, the goal is to search for and keep the local maximum NIR value found (to be saved as "DowelDistance1" later). The algorithm used is rather straightforward:

- We'll keep up with two "consecutive" and "running" NIR Sensor readings named "PreviousNIR" and "CurrentNIR".

- If "CurrentNIR" is greater than "PreviousNIR", we have not found the local maximum value yet, i.e. the "physical" centerline of the dowel is not yet found – well a good estimate of it any how!

- When "CurrentNIR" is less than "PreviousNIR" for the first time, we just pass over the local maximum, so we are going to save the value of "PreviousNIR" as "DowelDistance1" (or "DowelDistance2" depending on when one uses Function "FindDowelA").

```
101 FUNCTION FindDowelA
102 {
103     WAIT WHILE ( PORT[4]:IR Sensor <= DowelThreshold )
104     PreviousNIR = BackgroundNIR
105     CurrentNIR = PORT[4]:IR Sensor
106     NIRDifference = CurrentNIR - PreviousNIR
107     LOOP WHILE ( NIRDifference >= 0 )
108     {
109         PreviousNIR = CurrentNIR
110         Print Screen = PreviousNIR
111         CurrentNIR = PORT[4]:IR Sensor
112         Print Screen with Line = CurrentNIR
113         NIRDifference = CurrentNIR - PreviousNIR
114     }
115     Print Screen = PORT[3]:Servo Motor Position
116     Print Screen = PreviousNIR
117     FoundDowel = FoundDowel + 1
118     Print Screen with Line = FoundDowel
119 }
```

Fig. 3.59 Function "FindDowelA" in "DowelScanner.tskx".

Let's go through Fig. 3.59 to explain this algorithm line by line:

- First, we need to "seed" the parameters "PreviousNIR" and "CurrentNIR" with "logical" values. For "PreviousNIR", "BackgroundNIR" is definitely a logical one to use (Line 104), because at this time, we know that the NIR Sensor starts to "see" the edge of the dowel so its values would be for sure greater than "BackgroundNIR". For "CurrentNIR", it is of course logical to assign the latest reading from the NIR Sensor to it, so Line 105 is used. "NIRDifference" is computed next (Line 106), and **we know for sure that the WHILE LOOP on Line 107 would be entered next because of our previous choices for "PreviousNIR" and "CurrentNIR"**.

- The actions defined between Lines 109 and 113 are designed to swap the "CurrentNIR" value into "PreviousNIR" (Line 109) and print it (Line 110), next "CurrentNIR" is updated with the latest reading from Port 4 and it is also printed out. Finally, "NIRDifference" is updated and the loop's condition (NIRDifference > 0) is checked, and another iteration would be started if needed.

- Eventually, "NIRDifference" becomes negative when the dowel's centerline gets passed over, and the loop exits into Line 115 to print out the current Servo Position (for the user's benefit only), and then "PreviousNIR" gets printed. Finally, "FoundDowel" gets incremented by "1" and printed. When Function "FindDowelA" gets called for the second time by Function ServoScan on Line 93 in Fig. 3.58, the previous steps are executed and "FoundDowel" gets incremented to "2" and printed.

- There is a **subtle point regarding the Function "FindDowelA"** that the reader needs to be aware of. The PRINT commands in Lines 110 and 112 of Fig. 3.59 seem to be used there just to help the user monitor the progress of the algorithm, and for "full speed" performance the user may want to "comment" them out. However, if so done, the user will find out that the PORT 4 NIR Sensor readings are no longer correct and tend to be smaller in numerical values than they really are. Essentially, the process for sensor data acquisition may be a little slower than the speed that one's algorithm is capable of, thus creating some synchronization problems. There is another more robust approach using Function CALLBACK which will be described later in Section 3.9.3.

5) Once the previous Function "ScanServo" is done (Line 37 of Fig. 3.57), Parameters "DowelDistance1" and "DowelDistance2" are assigned with their actual values, then Function "VerifyDowels" is started (Line 38 of Fig. 3.57). Fig. 3.60 describes the details of Function "VerifyDowels" which is essentially an appropriate sequence of Function Calls with Functions "CenterSensor", "Right", "Left", and "FindApproachDowel". For example, if "DowelDistance2" is larger than "DowelDistance1" (Line 123), this means that the Right Dowel is closer to the robot than the Left Dowel, thus the robot first straightens up the NIR Sensor forward (Line 125) and then turns right towards the Right Dowel. Once the robot finds the Right Dowel with the NIR Sensor, it approaches this dowel and when it is close enough, it will play Melody 3 signaling to the user to remove the Right Dowel (Line 128). Next, it swings to the Left looking for the Left Dowel (Line 130) and performs similar tasks of finding and approaching the Left Dowel (Line 131). If the other situation happens instead (Line 133), i.e. the Left Dowel is closer to the robot than the Right Dowel, then the robot would execute similar find-approach tasks for the Left Dowel first (Lines 135-138) then turns right to take care of the Right Dowel (Lines 140-141).

```
121 FUNCTION VerifyDowels
122 {
123     IF ( DowelDistance2 > DowelDistance1 )
124     {
125         CALL CenterSensor
126         // Approach closer dowel on right
127         CALL Right
128         CALL FindApproachDowel
129         // Approach farther dowel on left
130         CALL Left
131         CALL FindApproachDowel
132     }
133     ELSE
134     {
135         CALL CenterSensor
136         // Approach closer dowel on left
137         CALL Left
138         CALL FindApproachDowel
139         // Approach farther dowel on right
140         CALL Right
141         CALL FindApproachDowel
142     }
143 }
```

Fig. 3.60 Function "VerifyDowels" in "DowelScanner.tskx".

The detailed codes for Functions "CenterSensor" and "FindApproachDowel" are listed below:

a) Function "CenterSensor" is quite simple (see Fig. 3.61). It just sets Servo Speed to 512 and Servo Position to Position 512, and then waits for 0.5 second for the Servo to get there.

b) Function "FindApproachDowel" (see Fig. 3.62) uses the same search strategy employed in Function "FindDowelA" to find the needed Dowel, but this time by rotating the whole robot Left or Right as needed using the Geared Motors on Ports 1 and 2 (i.e. the various Functions "Left", "Right", "Forward" and "Stop" using Parameters "SpeedL" and "SpeedR" previously set in the INIT code section). When "FindApproachDowel" was called from inside "VerifyDowels", the robot was already turning Left or Right appropriately to get to its targeted Dowel (Lines 127, or 130, or 137, or 140 in Fig. 3.60). Thus, the first statement inside "FindApproachDowel" is a WAIT WHILE (Line 147), active as long as the NIR Sensor reads in a value that is less than "DowelThreshold", i.e. the controller code is letting the robot scan "empty space" until the NIR Sensor starts "seeing" the Dowel's edge, resulting in readings larger than "DowelThreshold".

```
FUNCTION CenterSensor
{
    PORT[3]:Servo Motor Speed = 512
    PORT[3]:Servo Motor Position = 512
    High-resolution Timer = 0.500sec
    WAIT WHILE ( High-resolution Timer > 0.000sec )
}
```

Fig. 3.61 Function “CenterSensor” in “DowelScanner.tskx”.

```
FUNCTION FindApproachDowel
{
    WAIT WHILE ( PORT[4]:IR Sensor <= DowelThreshold )
    PreviousNIR = BackgroundNIR
    CurrentNIR = PORT[4]:IR Sensor
    NIRDifference = CurrentNIR - PreviousNIR
    LOOP WHILE ( NIRDifference >= 0 )
    {
        PreviousNIR = CurrentNIR
        Print Screen = PreviousNIR
        CurrentNIR = PORT[4]:IR Sensor
        Print Screen with Line = CurrentNIR
        NIRDifference = CurrentNIR - PreviousNIR
    }
    CALL Stop
    ProximityLimit = PORT[4]:IR Sensor
    ProximityLimit = ProximityLimit + 150
    CALL Forward
    WAIT WHILE ( PORT[4]:IR Sensor < ProximityLimit )
    CALL Stop
    Buzzer Timer = Melody Time
    Buzzer Index = Melody No.3 (3)
    WAIT WHILE ( Buzzer Timer > 0.0sec )
}
```

Fig. 3.62 Function “FindApproachDowel” in “DowelScanner.tskx”.

At that point, the controller quits the WAIT WHILE and starts on the search for the local maximum value of the NIR Sensor (Lines 148-158). This search strategy is already explained in the previous paragraph **4)**. Once this search is finished (Line 158), CALL STOP (Line 159) is issued to the robot as it should be now facing the Targeted Dowel squarely - well, most of the time in real practice, as it is less precise to stop the Geared Motors with the inertia of the whole robot to deal with, than to stop the much lighter NIR Sensor device as before (*if needed, try to insert a WAIT WHILE Timer Loop for about 0.25 second before the CALL STOP on Line 159)*. Next, the current value of the NIR Sensor is read in and its value saved in "ProximityLimit" (Line 160). Line 161 increments "ProximityLimit" by 150 units and this new "ProximityLimit" value is used later (Line 163) to tell the robot when to stop its approach towards the targeted dowel (Lines 162-164). The robot then plays Melody 3 to convey "Mission Accomplished"!

Video 3.28 shows a run-time demonstration of this project "DowelScanner.tskx" in its first half.

3.9.3 Using Function CALLBACK for Data Acquisition

The TASK tool also has an advanced Function named "CALLBACK" which cannot be not called from inside the MAIN or REGULAR Functions that the readers have seen so far. CALLBACK is called via the CM-150 hardware timer interrupt every 8 ms, independent of the current activities inside the MAIN or REGULAR Functions. It is restricted in size to only 512 bytes (i.e. an upper limit of 60-64 statements), and **only up to two device calls can be used per CALLBACK's instantiation** (see Fig. 3.63).

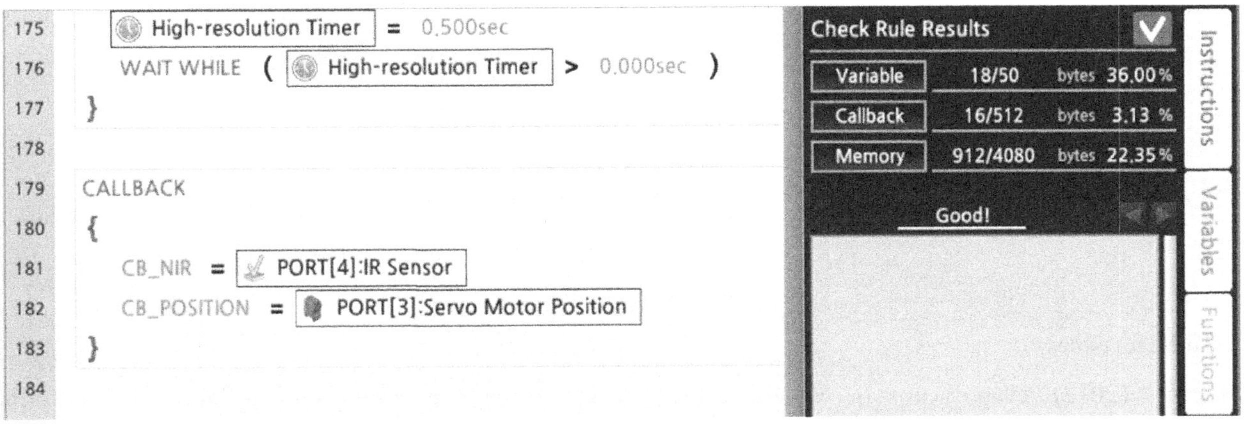

Fig. 3.63 CALLBACK Function used in "DowelScanner_CB.tskx".

Fig. 3.63 shows that for every 8 ms CALLBACK collects the current values of the Port 4 NIR Sensor and the Port 3 Servo Motor Position and saves them respectively in Parameters "CB_NIR" and "CB_POSITION". Consequently, the codes from inside the MAIN and REGULAR Functions need to be changed accordingly: i.e. each instance of "Port 4 NIR Sensor" is replaced by "CB_NIR", and each instance of "Port 3 Servo Motor Position" is replaced by "CB_POSITION", but ONLY when it is used as a READ parameter (i.e. when it is used on the RHS of an Assignment Statement or inside a Conditional Statement). Please compare Fig. 3.59 to Fig. 3.64 to see the appropriate changes made to Function "FindDowelA". The side benefit from using the CALLBACK function is that the PRINT statements are no longer needed in the LOOP WHILE code and that the NIR Sensor readings will be in the correct range of values. Please watch the second half of Video 3.28 for a run-time performance of "DowelScanner_CB.tskx".

```
FUNCTION FindDowelA
{
    WAIT WHILE ( CB_NIR <= DowelThreshold )
    PreviousNIR = BackgroundNIR
    CurrentNIR = CB_NIR
    NIRDifference = CurrentNIR - PreviousNIR
    LOOP WHILE ( NIRDifference >= 0 )
    {
        PreviousNIR = CurrentNIR
        // Print Screen = PreviousNIR
        CurrentNIR = CB_NIR
        // Print Screen with Line = CurrentNIR
        NIRDifference = CurrentNIR - PreviousNIR
    }
    Print Screen = CB_POSITION
    Print Screen = PreviousNIR
    FoundDowel = FoundDowel + 1
    Print Screen with Line = FoundDowel
}
```

Fig. 3.64 New Function "FindDowelA" using CALLBACK Parameters CB_NIR and CB_POSITION in "DowelScanner_CB.tskx".

3.10 References

Bergin J (2012) Pedagogical patterns: advice for educators. CreateSpace Independent Publishing Platform, p. 115.

Thai CN (2017) Exploring Robotics with ROBOTIS Systems (2nd Edition). Springer Publishers International, New York.

Thai CN (2018) Advanced Application Programming with ROBOTIS DREAM System. Amazon Kindle Direct Publishing.

3.11 Review Questions for Chapter 3

1. How often is the CALLBACK Function executed during runtime of a TASK program?
2. List the bit-level operators available in the TASK tool.
3. Does the CM-50 support floating-point arithmetics?
4. List 3 types of LOOPS in TASK.

5. From which side of a LOAD statement, does it begin to execute?

6. Is there any difference between the operators "=" and "=="?

7. How many Musical Instruments are allowed in the R+m.PLAY700 App?

8. How many Audio Sources are available to use in the R+m.PLAY700 App?

9. Describe the SMART CONSTANT structure used for Musical Instruments in the R+m.PLAY700 App.

10. Describe 3 programming techniques that can be used for programming robot autonomous behaviors.

11. Describe 3 ways to achieve Remote Control of robots.

12. Try to play a Music Melody without using WAIT WHILE (Buzzer Timer > 0). Does the music still play?

13. Use the Hi-Res Timer in the programs "IR-based-Maneuvers-IFs.tskx" and "IR-based-Maneuvers-IF-ELSE-IF.tskx" to figure out how many milliseconds does it take for the CM-150 to go through one iteration of its ENDLESS LOOP. This is to figure out which approach, Parallel IFs or IF-ELSE-IF, is more time efficient.

14. Instead of the "cumulative" values, try "constant" values for the Steering Angle and Speed Differential parameters in the program "RC-TriCycle.tskx". What is the effect on the driving behaviors of the TriCycle?

15. Among the accompanying example codes, the reader can look for and try out another Remote-Control approach whereas the L and R buttons are to be "pulsed" (with audio feedback) for the Left and Right turns to work properly (see "RC-TriCycle-LR_Pulse.tskx"). Furthermore, in this scheme, the buttons U-D, L-R and 1-3 are treated as independent button groups.

16. Create a new Remote-Control scheme for the TriCycle robot where the powered wheels are considered as the front of the vehicle and the servo-controlled steering wheel is considered as the rear of the vehicle (also known as the "Tadpole" configuration).

17. Based on the Mixed-Control approaches shown in the programs "RC-TriCycle-LR_Pulse.tskx" and "MotoCycle.tskx" (see Chapter 6), please create a new TSKX program that can use the NIR Sensor (Port 4), mounted on the steering wheel (see Fig. 3.44), to allow the TriCycle to avoid a frontal obstacle.

18. In the example code "SmartAvoider.tskx", try to replace lines 111-115 with WAIT WHILE (Recon Data Received == TRUE). Does the robot still behave properly?

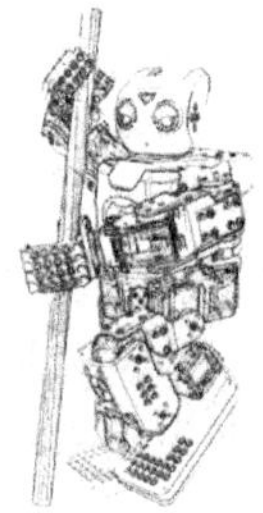

Chapter 4: Using R+SCRATCH & SCRATCH® 2

MIT's SCRATCH 2 software is a very popular software platform with millions of users worldwide (https://scratch.mit.edu/), and reference texts on this subject are too numerous to list in this chapter. The author would recommend such works as Ford (2014), Warner (2015) or Vlieg (2016) for self-learners who want to have a thorough understanding of the SCRATCH 2 language. If the reader is a teacher or trainer then the author would also recommend the book by Bagge (2015) and his Code-IT web site (http://code-it.co.uk/philbagge). In this Chapter, only selected SCRATCH 2 features, most applicable to robotics, would be presented.

For the CM-150, ROBOTIS designs the software tool R+SCRATCH, **V.1.0.9 or above**, to be a helper application that links the SCRATCH 2 IDE directly to the CM-150 Controller's Firmware, **V.31 or above** (see Figs. 1.8 and 4.1). This means that any TASK code previously downloaded onto the CM-150 is disabled (not erased) as soon as R+SCRATCH connects to the CM-150 (see Part 1 of Video 4.1).

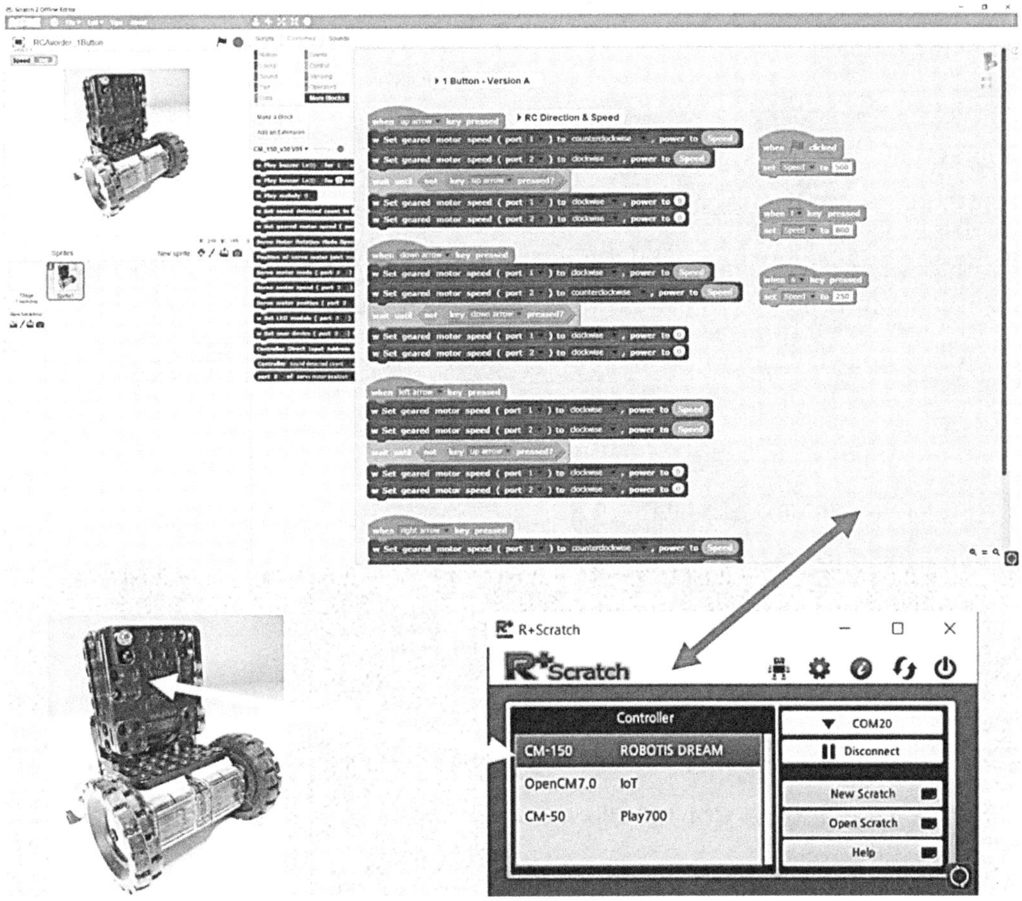

Fig. 4.1 Information Flow between SCRATCH 2, R+SCRATCH and CM-150 Controller.

ROBOTIS also creates a custom SCRATCH HTTP Extension tool (https://wiki.scratch.mit.edu/wiki/Scratch_Extension) to work with the SCRATCH 2 Off-Line Editor to make available 15 SCRATCH Block Codes that can interface with the CM-150's firmware (see Fig. 4.2). The top 13 blocks are of the "WRITE" type of command. Each "WRITE" command requires 50-70 ms delay to progress, from the SCRATCH 2 IDE through the R+SCRATCH helper application, and all the way down to the CM-150 Firmware and actuators mounted on the CM-150 robot. The "Set Value of Address" WRITE block is for advanced users as it will require referencing back to the Control Table of the CM-150 (http://emanual.robotis.com/docs/en/parts/controller/cm-150/). There are only two "READ" type of command, "Controller Value" and "Port Value" which have a pull-down menu to access various devices, internal or mounted onto Ports 3 or 4 (see Fig. 4.3). Each "READ" command requires 120-150 ms delay to process, once it is triggered at the SCRATCH 2 IDE level. When using TASK codes as shown in Chapter 3, these types of commands cost much less time to be performed (less than 5 ms), so the much longer time delays when using SCRATCH 2 create important and sometimes adverse side effects (programming and physical) affecting a robot runtime performance. These effects will be shown and discussed further in later parts of this Chapter and in Section 6.7.

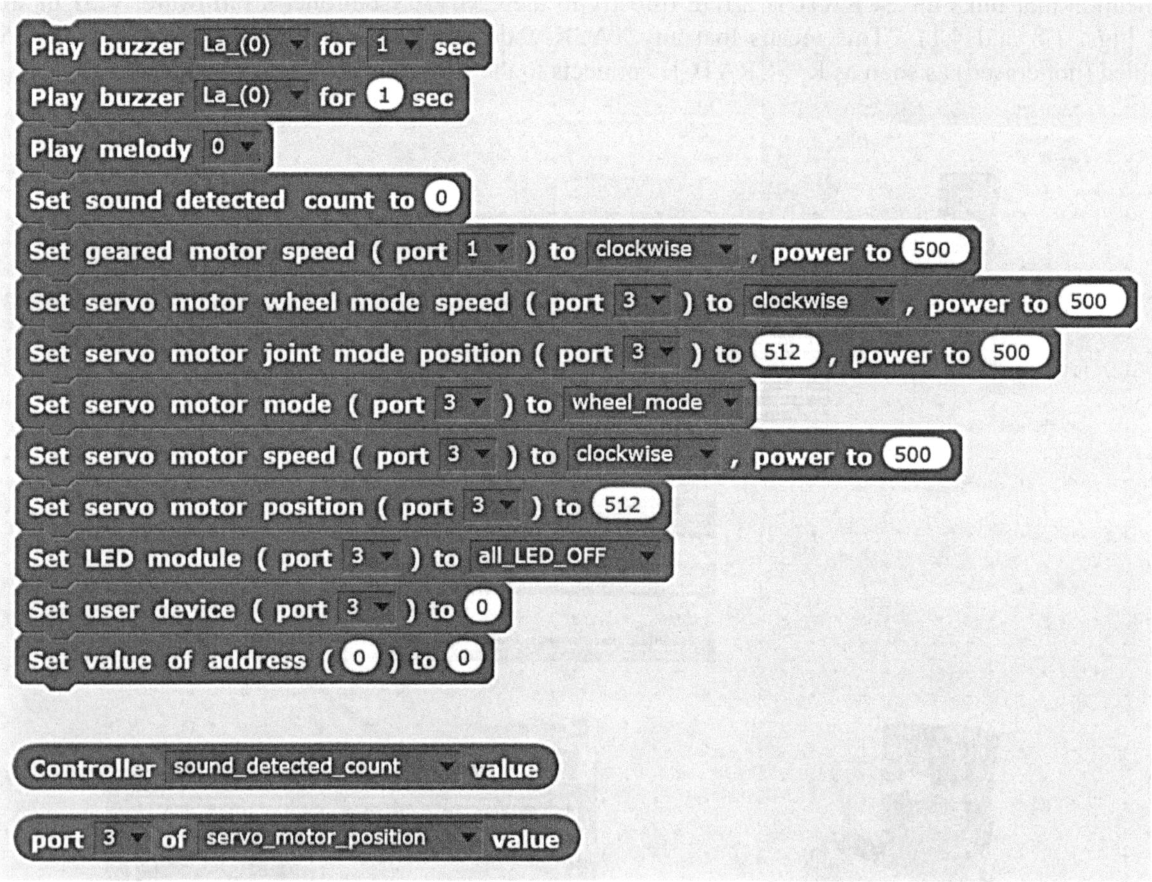

Fig. 4.2 ROBOTIS CM-150 Block Codes compatible with SCRATCH 2 IDE.

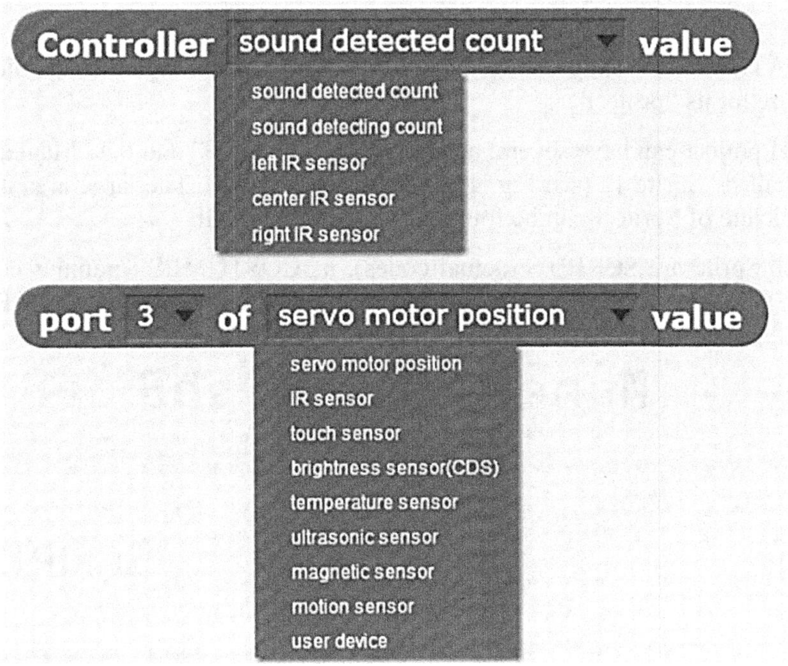

Fig. 4.3 Pulldown Menu Options for "Controller Value" and "Port Value" blocks.

Unfortunately, please note that currently the Color Sensor (CS-10) and the Temperature/Humidity Sensor (TMS-10) cannot be ported to work with SCRATCH 2 for technical reasons.

Part 2 of Video 4.1 shows a basic usage of the tool chain R+SCRATCH/SCRATCH 2 Offline Editor and the "FlashDancer" robot (see Fig. 4.4), resulting in a project named "Maneuvers-1.sb2" to be described in Section 4.1 (see Fig. 4.5).

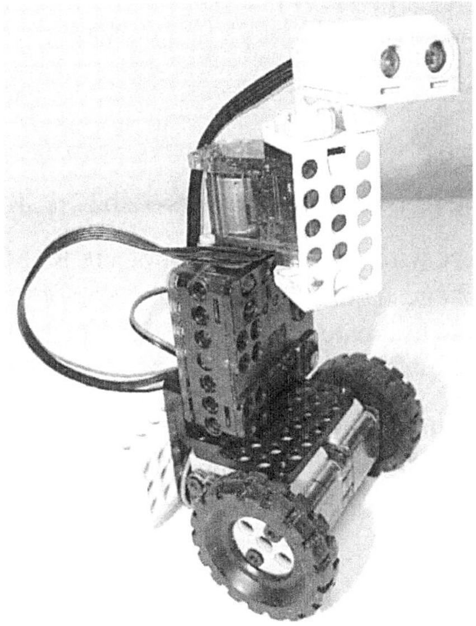

Fig. 4.4 "FlashDancer" robot.

Although SCRATCH 2 is designed for beginner programmers aged 8 or older, it has quite a sophisticated structure for its "project":

Each SCRATCH project can have several actors called "SPRITES" and the "Maneuvers-1" project uses a single Sprite called "Sprite 1" (see Fig. 4.5 – bottom left panel). The large area on the top left panel, where a larger picture of Sprite 1 can be found, is called the STAGE.

Attached to each Sprite are SCRIPTS (actual codes), its COSTUMES (graphics components triggered by its SCRIPTS), and its SOUNDS (audio components also controlled by its SCRIPTS) – see top of mid-panels of Fig. 4.5.

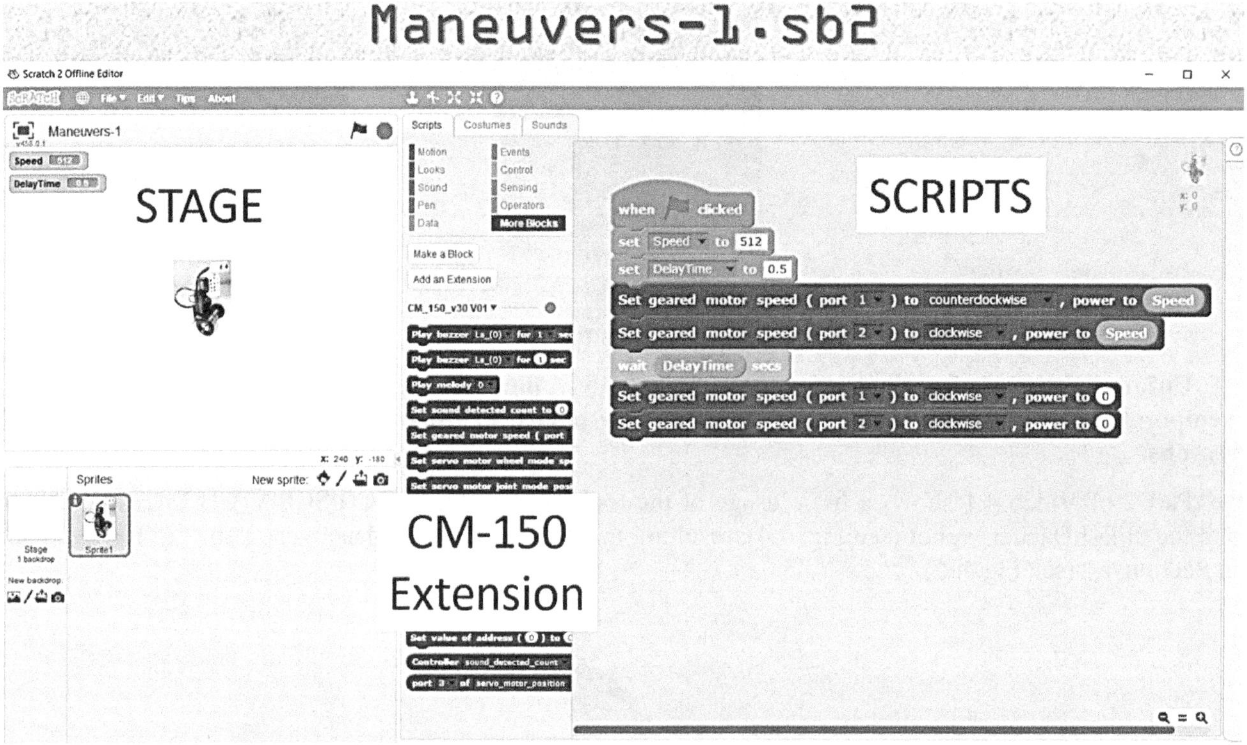

Fig. 4.5 Components of the "Maneuvers-1.sb2" SCRATCH 2 project.

To allow SCRATCH 2 to interact with the CM-150 robot via R+SCRATCH, the user must also mount appropriate "EXTENSION" packages (i.e. a *.JSON file – see bottom of mid-panel in Fig. 4.5). Please see Video 4.2 for a primer on how to create and edit a SCRATCH 2 project that would interact with this example robot.

The right panel contained the actual project's SCRIPTS which were created using different types of colored blocks (representing different data and behavior structures) that the user could stack and unstack to build the logic required for a project.

4.1 Basic Robot Programming with SCRATCH

In this section, the goal is to show the similarities and differences in creating and running 2 pieces of codes on the same robot with the same required physical behavior (i.e. "go forward for 0.5 second"). One is a TASK code from Section 3.2 named "Maneuvers-1.tskx" and the other is a SCRATCH program called "Maneuvers-1.sb2" (see Video 4.3 for complete steps).

Fig. 4.6 displays the MAIN SCRIPT which starts when the user clicks on the GREEN FLAG block (which has similar functions as the START PROGRAM statement used in TASK codes):

The GREEN FLAG block can be found in the EVENTS menu located in the SCRIPTS tab.

Next, Variables "Speed" and "DelayTime" are set respectively to "512" and "0.5" (lines 1-2, i.e. same functionality as Assignment statements inside TASK). The Variable parameters can be created and set using items found in the DATA menu of the SCRIPTS tab.

Then, Port 1 Motor is set to a Counter-Clockwise motion with a power equaled to Variable Speed (line 4), while Port 2 Motor is set to a Clockwise motion with a power also equaled to Speed (line 5). Together, these two Code Blocks make the Bot go forward. Inside TASK codes, we would use an equivalent COMPUTE type of command here. These Code Blocks are part of the ROBOTIS' EXTENSION so they can be found in the MORE BLOCKS menu of the SCRIPTS tab.

Line 6 just makes the PC (not the actual robot) wait for 0.5 second, because a SCRATCH program runs inside the PC's memory and **not** on the CM-150's memory like for a TASK code. The WAIT block can be found in the CONTROL menu of the SCRIPTS tab. This step is clearly easier to use than setting up the Hi-Res Timer with a delay of 500 ms in TASK codes as shown in Section 3.2.

Lastly, lines 7 and 8 turn off the motors to make the robot stop.

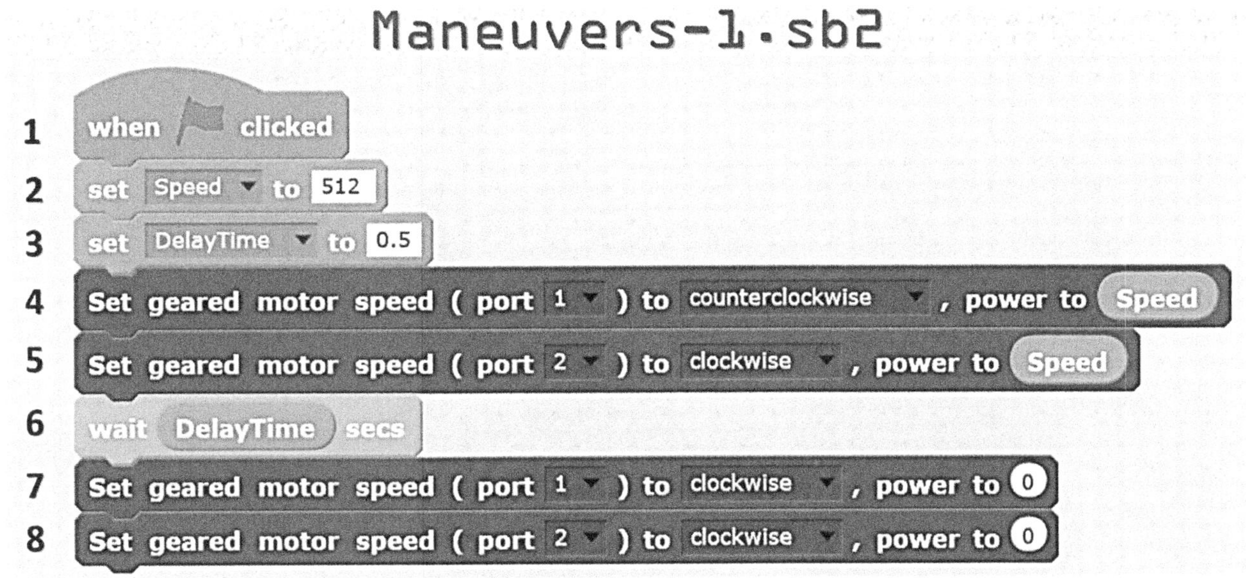

Fig. 4.6 MAIN SCRIPT for "Maneuvers-1.sb2" project.

Video 4.3 also contrasts the runtime performances of this robot when running a SCRATCH code (Maneuvers-1.sb2) and when running an equivalent TASK code (Maneuvers-1.tskx). For the SCRATCH version, the subtle "jerks" at the beginning and end of the robot's forward motion are due to the short delays for a SCRATCH command to go from the PC to the robot Controller to activate a given motor. Essentially, Port 1 Motor (on robot's left) was activated/deactivated a few milliseconds before Port 2 Motor (on robot's right) and thus created the beginning "right" jerk and the final "left" jerk. Logic-wise, the same timing delays are present in the TASK code, but because it runs directly inside the CM-150 controller's memory space, and therefore these commands can reach the left and right motors much faster. This explains why these effects were not "obvious" when using the TASK tool in Chapter 3.

4.2 Sequential Control of Robot Maneuvers

When using TASK, back in Section 3.2, we learned how to create independent program modules called FUNCTIONs which were then had to be CALLed from inside the MAIN Task program to execute these functions (see Figs. 3.5 and 3.6). In SCRATCH, ones can create similar independent code-block modules as shown in "Maneuvers-3.sb2" (see Fig. 4.7). These SUB-SCRIPTs were duplicated from the single one found in "Maneuvers-1.sb2" (Fig. 4.6) and individually modified with the appropriate turn directions, clockwise or counter-clockwise, to create four robot maneuvers: Going Forward, Going Backward, Turn Left and Turn Right. To activate any SUB-SCRIPT, the user only needs to click on the appropriate GREEN FLAG block and the robot performs the user-wanted maneuver (see Video 4.4).

The author is quite certain that the reader already notices that this is a completely new and **interactive** way to think about problem-solving, code generation, execution and trouble-shooting when using SCRATCH, as compared to the longer cycle of "problem-solving/editing/compiling/downloading/executing/trouble-shooting" needed when using TASK. This is the first SCRATCH feature that promotes a "parallel/lateral thinking" mode for a SCRATCH practitioner.

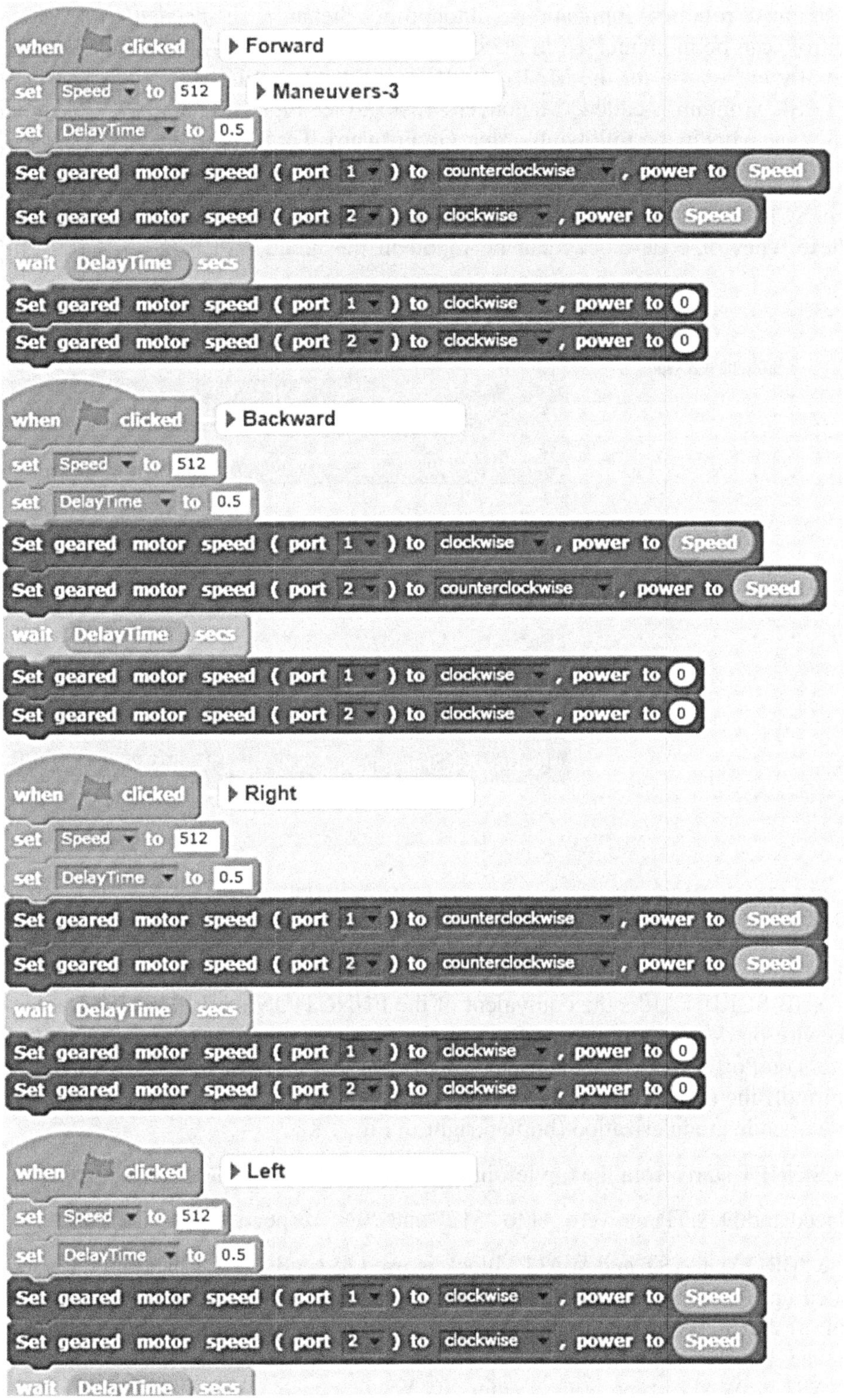

Fig. 4.7 Independent Maneuvers SUB-SCRIPTs for "Maneuvers-3.sb2" project.

However, in most robotics applications, autonomous behavior is needed, thus let's first see how "sequential control" can be implemented in SCRATCH. The resulting program is "Maneuvers-4a.sb2" and it is shown partially in Fig. 4.8 for the MAIN SCRIPT and the two SUB-SCRIPTs "Forward" and "Stop". The matching TASK program is called "Maneuvers-4.tskx" (see Figs. 3.8 and 3.9). The goal is to make the robot perform 3 maneuvers in the following order: Go Forward, Turn Right, Go Backward and then stops.

The project "Maneuvers-4a.sb2" (Fig. 4.8) implements another powerful "parallel-thinking" feature, available only in SCRATCH, which is the ability to BROADCAST and RECEIVE messages among various SCRIPTS. These types of code-blocks can be found in the menu EVENTS in the SCRIPTS tab of the SCRATCH IDE.

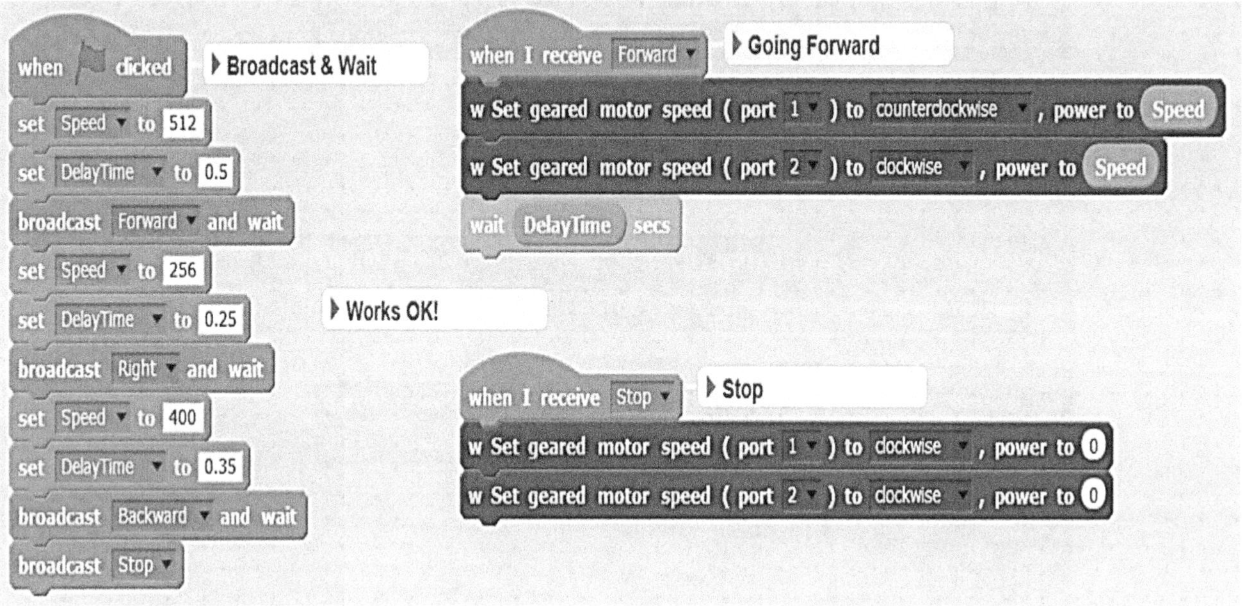

Fig. 4.8 MAIN SCRIPT and two Maneuvers SUB-SCRIPTs in "Maneuvers-4a.sb2" project.

Let's have a closer look at the SUB-SCRIPT "Forward" (top right in Fig. 4.8). The "When I Receive" type of block can be found in the menu "EVENTS" of the SCRIPTS tab and it should be the FIRST block for this type of SUB-SCRIPT. It is the equivalent of the FUNCTION's DEFINITION statement in a TASK code which likewise marks the beginning of a Function (see Section 3.2). Next are the two commands to turn on the motors on Port 1 and Port 2, then the "WAIT for DelayTime" block. However, the previous two commands to turn off the motors (see Fig. 4.7) are now set up in their own SUB-SCRIPT called "Stop" - as another example of code modularization (bottom right in Fig. 4.8).

The MAIN SCRIPT starts from the top left in Fig. 4.8 with the GREEN FLAG block:

1) First, Speed and DelayTime were set to "512" and "0.5" respectively.

2) Then the "BROADCAST and WAIT" block is used to send out the message "Forward" which is of course picked up by the SUB-SCRIPT "Forward". Here there are 2 parallel and simultaneous actions that happen: a) the SUB-SCRIPT "Forward" executes its blocks which makes the robot go forward at speed 512 and for 0.5 second; b) the MAIN SCRIPT is still "live" but as its current block is a "BROADCAST & WAIT" type, it now enters its WAIT phase. When the SUB-SCRIPT "Forward" is done, this SUB-SCRIPT terminates itself, then the MAIN SCRIPT can go on to the next block, which is to set the Variable Speed to a new value of "256". These steps are comparable to the ones taken by a

TASK program when it is calling a Function (section 3.2), however there is only **one** program flow-control pointer used in a TASK code, while SCRATCH can keep track of "multiple" program flow-control pointers (i.e. multiple BROADCAST messages, all thanks to the PC's powerful multi-tasking Operating System).

3) After setting Speed and DelayTime to their new values of "256" and "0.25", the MAIN SCRIPT issues another "BROADCAST & WAIT" block, this time to the SUB-SCRIPT "Right" which makes the robot turns right at speed 256 and for 0.25 second.

4) Next, the MAIN SCRIPT sets the new values of "400" and "0.35" respectively to "Speed" and "DelayTime".

5) The third maneuver "Going Backwards", at speed 400 and for 0.35 second, is initiated by the third "BROADCAST & WAIT" block.

6) Finally, to stop the bot, a plain "BROADCAST" command is used as it is at the end of the MAIN SCRIPT. Here, the reader can also use "BROADCAST & WAIT" with no adverse consequence to the behavior of the SCRATCH code as well as the behavior of the robot.

Video 4.5 shows the runtime performances of "Maneuvers-4.tskx" and "Maneuvers-4a.sb2" programs, with the usual "jerks" side effects occurring for "Maneuvers-4a.sb2". Video 4.5 also shows that the program "Maneuvers-4b.sb2", where only "BROADCAST" blocks are used, does not perform as expected (see Fig. 4.9). Essentially, the robot was "hit" with 3 commands "Forward", "Right" and "Backwards" practically at the same time, and it tried to implement them all which made it look like that it was "paralyzed", from the user's point of view.

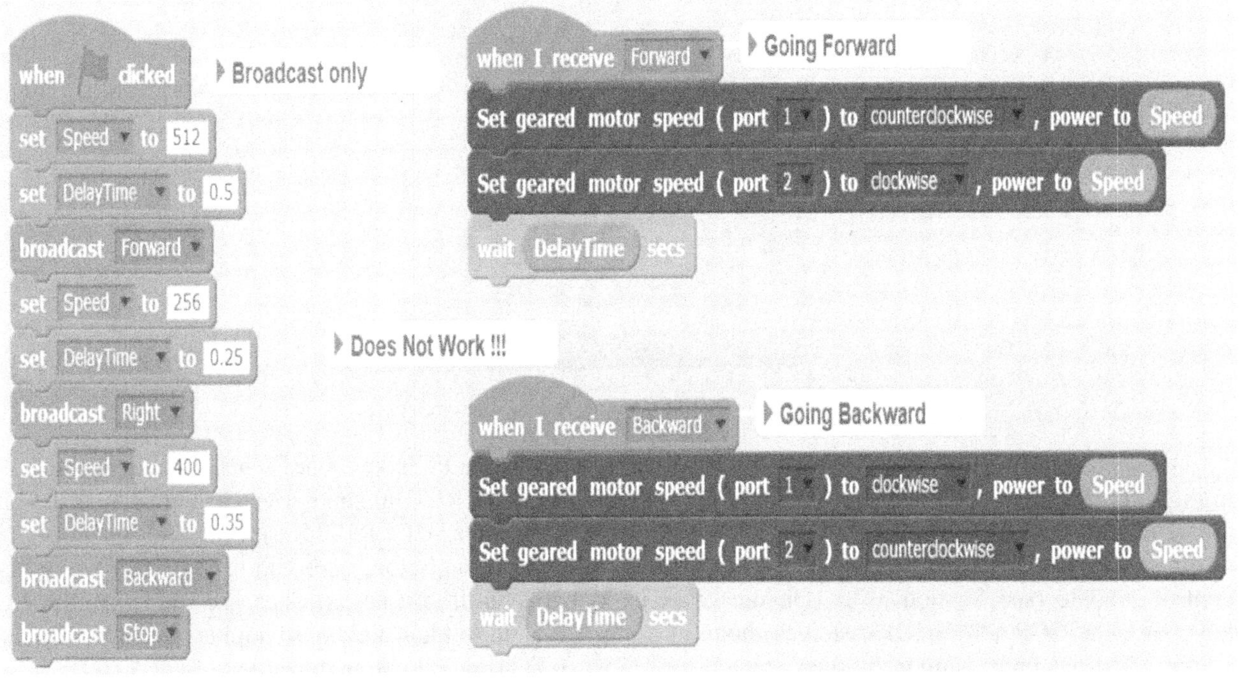

Fig. 4.9 MAIN SCRIPT and two Maneuvers SUB-SCRIPTs in "Maneuvers-4b.sb2" project.

4.3 Remote Control of "FlashDancer" via PC's keyboard

The SCRATCH/R+SCRATCH tool chain does not support a Virtual RC-100 Remote Controller like for the TASK tool (see Section 3.7), as SCRATCH already has a block that monitors the PC's keyboard, called "When xxx Key Pressed" and it is available in the EVENTS menu of the SCRIPTS tab (see Fig. 4.10).

Fig. 4.10 EVENTS Menu in SCRIPTS Tab

Another big difference between the RC-100's behavior and the PC's keyboard's behavior is in the way multiple buttons or keystrokes are handled. At any one time, the RC-100 sends over to the robot a 10-bit message that represents the status (OFF or ON) of all its 10 buttons (UDLR123456), thus the user needs to create special TASK codes to sort out specific data from each button before ones can use them in a remote-control scheme (see Section 3.7). On the other hand, when multiple key strokes are made on the PC keyboard, the SCRATCH IDE receives them as a stream of individual keyboard inputs, i.e. none of the communications processing techniques created for TASK in Section 3.7 is transferrable to SCRATCH.

In SCRATCH, the general approach is to create independent SUB-SCRIPTS to handle each key stroke of interest, and a MAIN SCRIPT to handle the interactions between keystrokes. The project "RC-FlashDancer-a.sb2" shows how to use the Up-Down-Left-Right Arrow keys for affecting the moving directions of "FlashDancer" (see Fig. 4.4), the keys "1" and "2" to switch respectively between a Slow

Speed and a Fast Speed regime, the key "f" to flash of the LED module 10 times and the key "s" to move the Servo Motor back and forth 3 times.

Fig. 4.11 shows only the SUB-SCRIPT handling the "Up-Arrow Key", when it is sent from the PC into the SCRATCH IDE (the other Arrow keys are handled similarly):

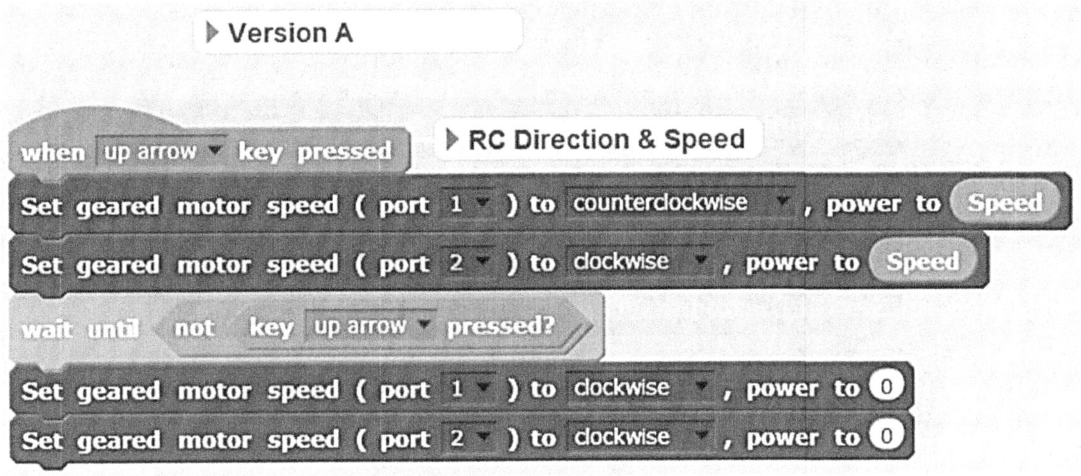

Fig. 4.11 SUB-SCRIPT handling the Up-Arrow Key.

When the Event "Up-Arrow Pressed" is detected, Ports 1 and 2 are set to a Power Level equal to the value of Variable Speed, with appropriate motor turn directions to make the "FlashDancer" go forward.

Next a WAIT UNTIL block (from the CONTROL Menu) is used to monitor when the user releases the Up-Arrow Key, i.e. when the Up-Arrow key is NOT PRESSED.

The last actions are to turn off power on the two motors to put the "FlashDancer" into a STOP.

Fig. 4.12 shows how the MAIN SCRIPT can get started to set Variable Speed to 250, and to get Port 3 Servo Motor to set Mode to "Joint", Speed to "1023" and Initial Position to "512". It also shows how the "2" key is monitored to set Speed to 800 and how the "1" key is monitored to set Speed to 250.

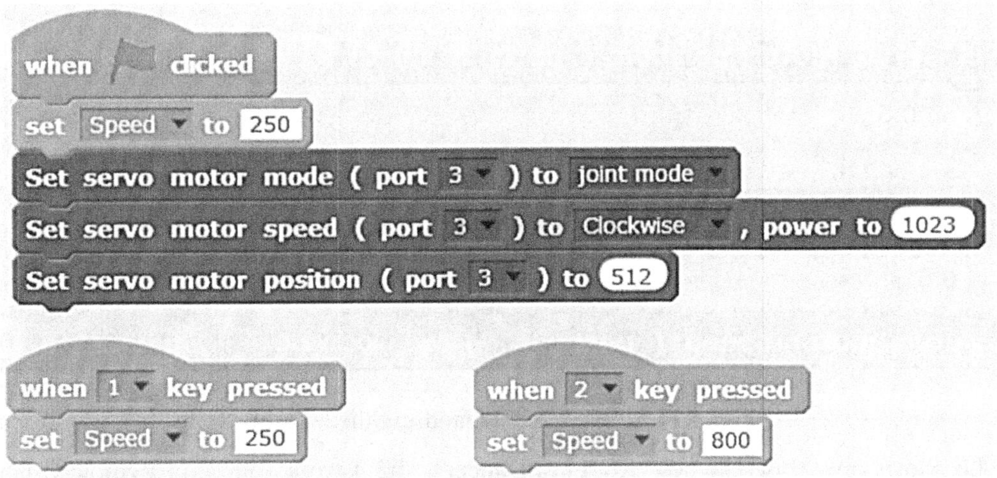

Fig. 4.12 MAIN SCRIPT and SUB-SCRIPTs handling the "1" and "2" Keys.

Fig. 4.13 shows how the "f" key is monitored to "flash" the LED module on Port 4 10 times and Fig. 4.14 show how the "s" key is monitored to "shake around" the Port 3 Servo Motor 3 times, to reset its final position to "512" and to set its final speed to zero (i.e. to turn it off to save battery).

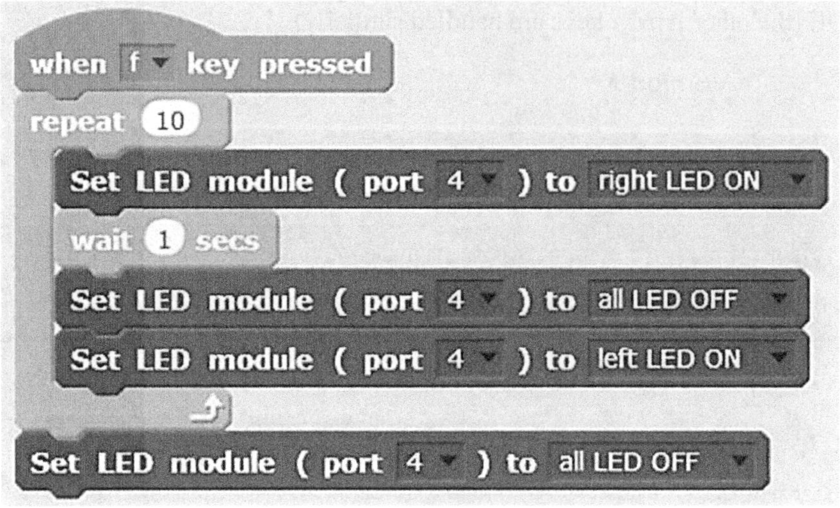

Fig. 4.13 SUB-SCRIPT handling the "f" Key.

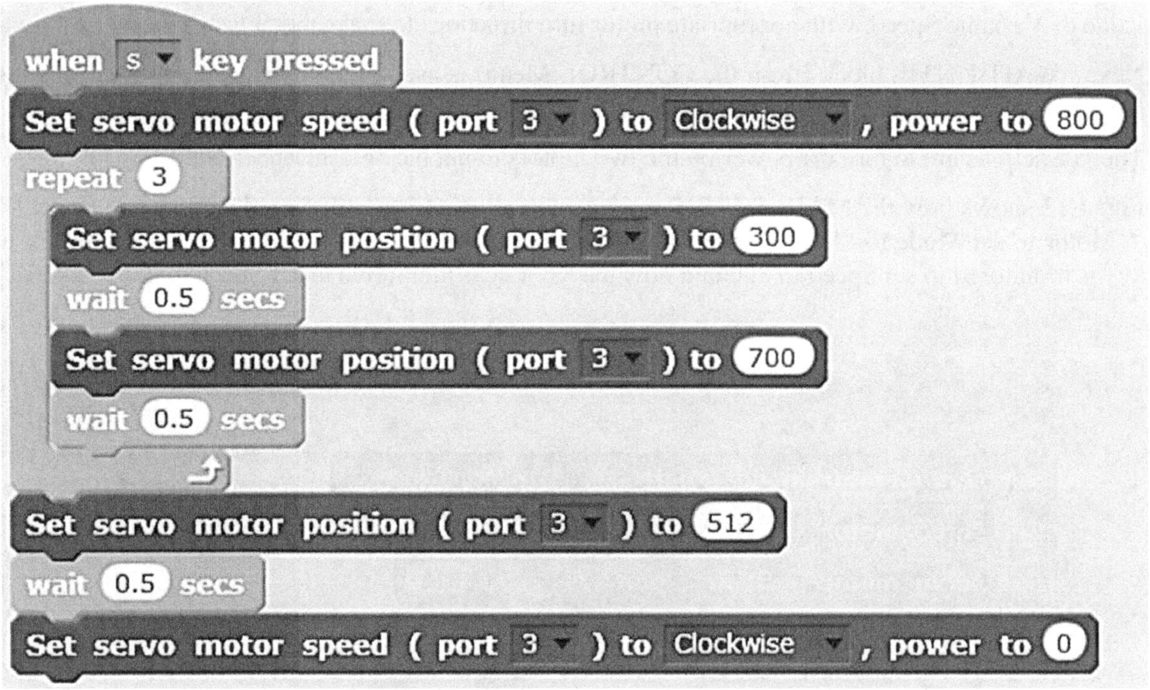

Fig. 4.14 SUB-SCRIPT handling the "s"Key.

Video 4.6 shows how this scheme ("RC-FlashDancer-a.sb2") works out as a Remote-Control scheme whereas the user can push down one key or multiple keys at each time. Video 4.6 shows a very responsive RC system.

The project "RC-FlashDancer-b.sb2" shows an alternate solution where the job for handling the release of the Arrow Keys is shifted into the MAIN SCRIPT (see Fig. 4.15):

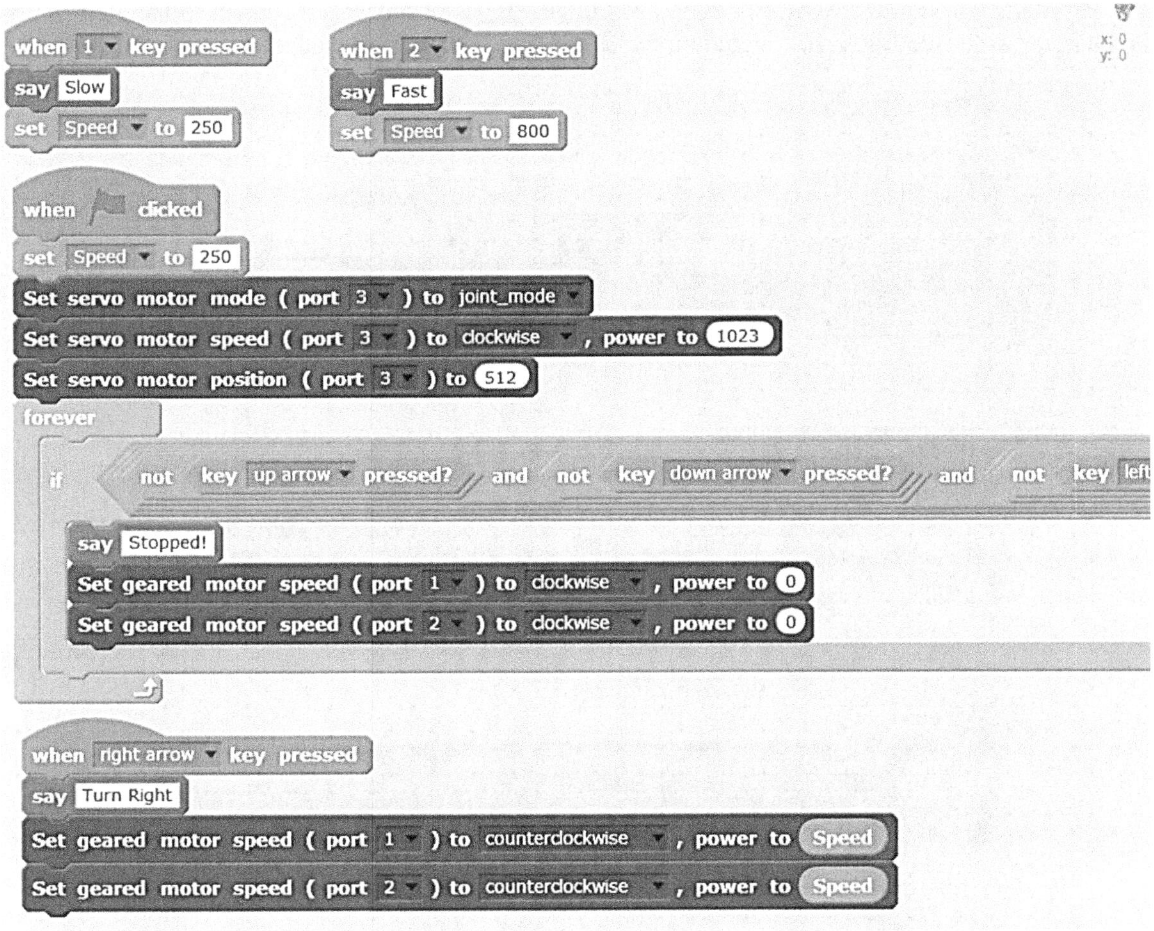

Fig. 4.15 MAIN SCRIPT and selected SUB-SCRIPTs in "RC-FlashDancer-b.sb2".

1) The MAIN SCRIPT starts out by SETting Variable Speed to 250 and other parameters for Port 3 Servo Motor as before, then it enters an FOREVER LOOP. Inside this loop, an IF block, with a rather complex "AND" Conditional Expression representing the situation when All of the Used Keys are NOT pressed, it is also used to SAY "Stopped" and to turn power off on both motors.

2) Fig 4.15 also shows "enhanced" SUB-SCRIPTS for key strokes "1" and "2" using appropriate Text Balloons.

3) The SUB-SCRIPT for "Right Arrow" is typical for all Arrow Keys, and it shows that these SUB-SCRIPTs are now only concerned with setting appropriate turn directions and power levels to the motors to make the robot go in the user-wanted directions.

Video 4.7 shows the run-time performance of "RC-FlashDancer-b.sb2". At the end of Video 4.7, the author shows that this approach of using multiple independent subscripts triggered by various keys did not perform as expected when the user wants to hold down for example Up and Right arrow keys to obtain a "Forward and Right" motion, similarly to the robot's behaviors obtained in the previous TASK codes "RC-MultiDirectionsSpeeds.tskx" and "RC-MultiDirectionsSpeeds-Mobile.tskx".

The project "RC-FlashDancer-Parallel-IFs.sb2" allows a combination of "directional" key strokes to be activated by the user: for example, "Up and Right" or "Down and Left". This capability is achieved by putting Parallel IFs inside the MAIN FOREVER LOOP. Each IF block is designed to handle a given key stroke being pressed (see Fig. 4.16), including the case when no key is pressed (see Fig. 4.17).

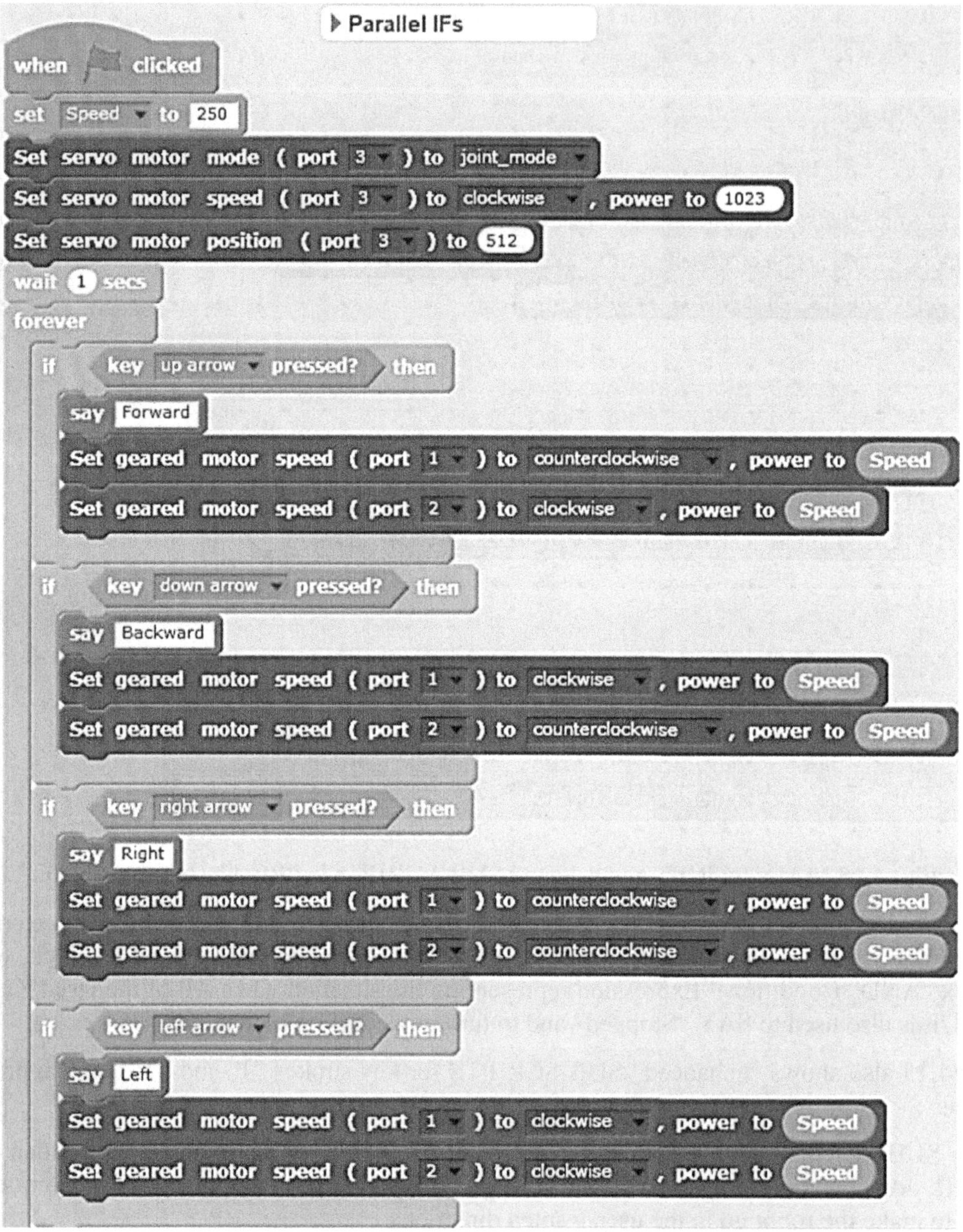

Fig. 4.16 Part 1 of MAIN SCRIPT in "RC-FlashDancer-Parallel-IFs.sb2".

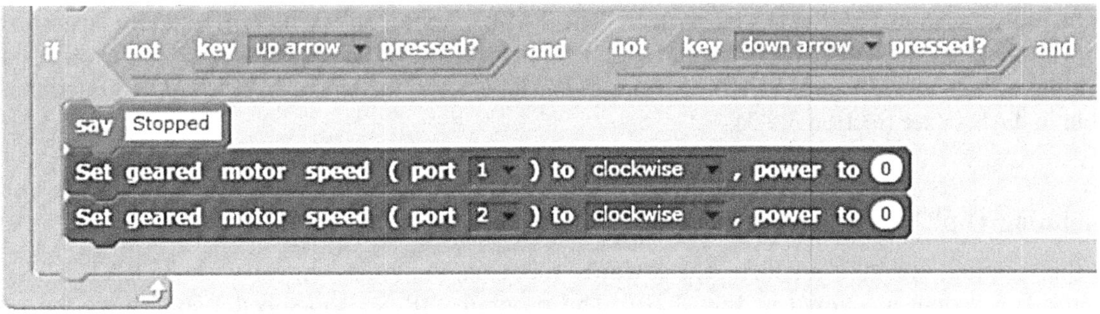

Fig. 4.17 Part 2 of MAIN SCRIPT in "RC-FlashDancer-Parallel-IFs.sb2".

The first part of Video 4.8 shows a runtime performance of the program "RC-FlashDancer-Parallel-IFs.sb2". The reader should compare this solution to the equivalent TASK solution "RC-MultiDirectionsSpeeds.tskx" (see Section 3.6.2 and Video 3.21) – a lot more work was needed in the TASK version! This first part of the video shows that the robot did try, for example, to do a proper sequence of "Forward" and "Right" motions, but that the "Forward" motion was too short timewise. Thus a "Wait 0.25 sec" blocks are inserted for the "Forward" and "Backward" codes (see Fig. 4.18) resulting in the behaviors shown in the second part of Video 4.8.

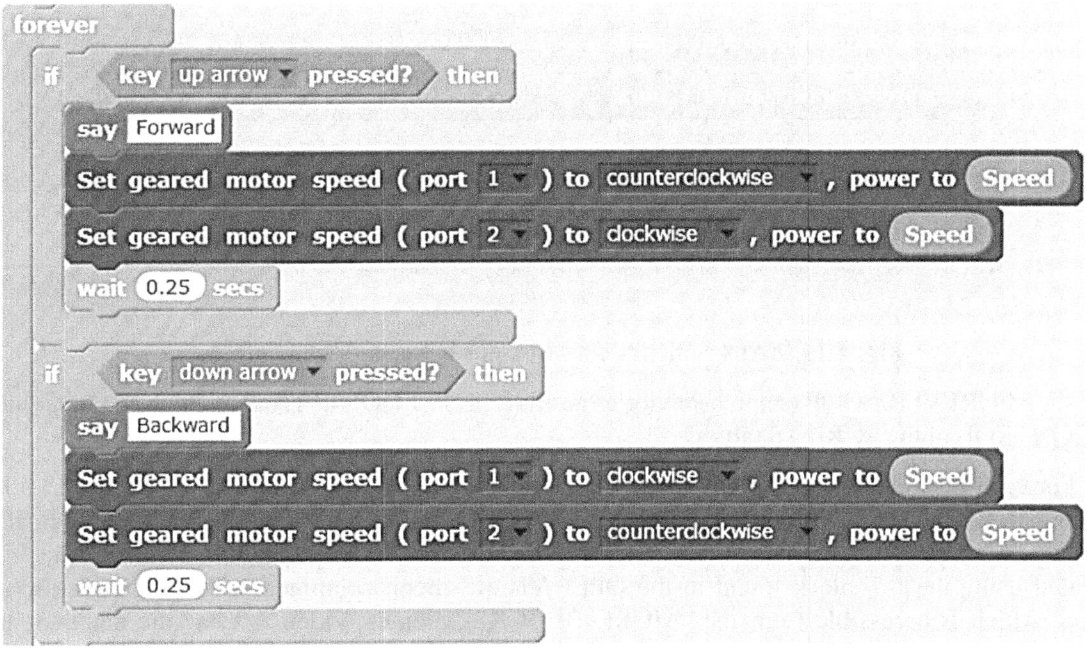

Fig. 4.18 Adding Wait Blocks in "RC-FlashDancer-Parallel-IFs-w.sb2".

In viewing Video 4.8, the observant reader may have noticed that the combined motions "D+R" and "D+L" as programmed in "RC-FlashDancer-Parallel-IFs-w.sb2" do not match with the same programmed motions in "RC-MultiDirectionsSpeeds-Mobile.tskx". After some further investigations, the author has found that SCRATCH 2.0 IDE does not handle properly simultaneous press-and-hold of multiple keys on the PC side (unlike the "Remocon" packet discussed in Section 3.7). For example, if the user presses and holds down on both "Up" and "Right" Arrow Keys, the typical SCRATCH code would receive 3 signals in a rolling sequence as "Up", "Right" and "Up & Right" (and not just "Up & Right" as in the referred TASK code). In other words, the robot behaviors as programmed in "RC-MultiDirectionsSpeeds-Mobile.tskx" cannot be reproduced in a SCRATCH 2 project. The TASK "Remocon" packet offers more customization possibilities than the SCRATCH IDE can provide for regarding keyboard processing. On the other hand, the multithreading capabilities of SCRATCH are much superior to the single CALLBACK Function available in TASK (see Section 3.9.3).

4.4 "Spinning Top" Maneuvers using Center IR Sensor

In this section, we take another step towards robot autonomous control with SCRATCH using the "Spinning Top" robot as shown in Fig. 3.10. The program "IR-Sensor-Speed-1.sb2" links the real-time value of the Center IR Sensor to the Forward motion of the robot (see Fig. 4.19). Not shown in this Fig. 4.19 are the definitions of the Variables "Speed" and "IR" via the DATA menu in the SCRIPTS tab.

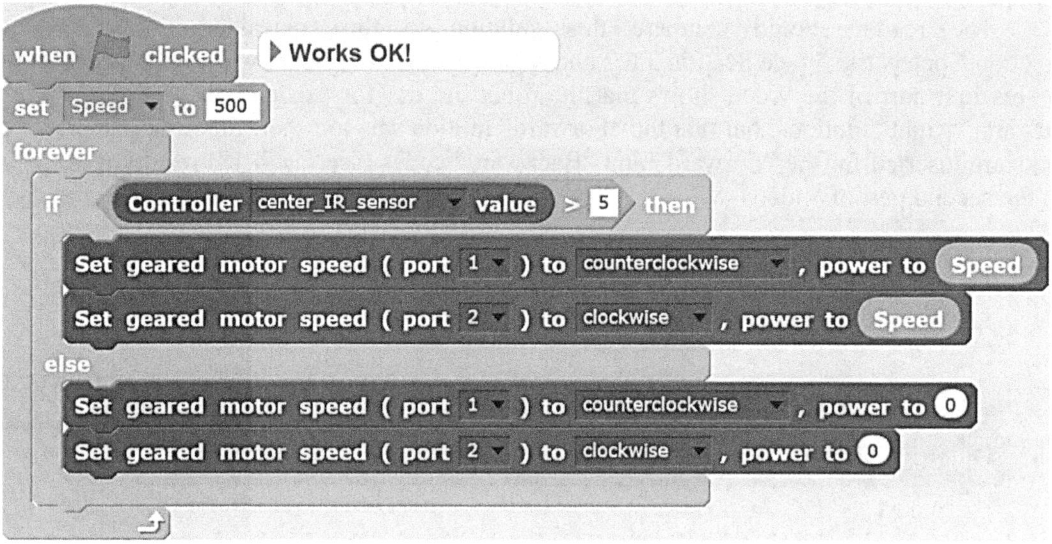

Fig. 4.19 MAIN SCRIPT for "IR-Sensor-Speed-1.sb2" project.

It uses a FOREVER LOOP (same behavior as an ENDLESS LOOP in TASK) which can be found in the menu CONTROL of the SCRIPTS tab:

1) For each iteration of this FOREVER LOOP, a READ "Controller Value" block is issued to the CM-150 Controller to return the current reading of its Center IR Sensor which is then used in an IF-ELSE structure (it can also be found in the CONTROL menu). The conditional expression of the IF part is created using the ">" block found in the OPERATORS menu combined with the "Controller Value" block which is accessible from the "MORE BLOCKS" menu. Video 4.9 has the detailed steps for creating such an IF-ELSE structure.

2) During any iteration when this conditional expression evaluates to TRUE, the robot is made to go forward at Speed 500. Otherwise, the robot stays still.

The runtime behavior of this program is also shown in Part 1 of Video 4.9: when the robot detects an object close enough its "head", it would dart forward to get away, then it would stop - ready to play again!

In the program "IR-Sensor-Speed-2.sb2", the value of the current Center IR Sensor is first saved into a SCRATCH parameter named "IR_Sensor" which is then used in the IF-ELSE structure. Part 2 of Video 4.9 shows that both programs behave the same way, but with "IR-Sensor-Speed-2.sb2" the user gets to see the actual value of the Center IR Sensor being displayed in the Stage area.

Now we are ready for our first implementation of the Sense-Think-Act paradigm previously discussed in Chapter 1 to the SCRATCH environment. The sensing component will be the IR Center Sensor of the "Spinning Top" robot as before.

The tasks assigned to this robot are as follows (review Section 3.3 also):

1) At start, the robot plays a musical melody (Melody 2) to let the user know that it is "ready for action", and its Speed initialized to zero.

2) Next, the behaviors of the robot depend on the position of an object relative to the IR Center Sensor, thus the Condition-Action list to be considered is as follows:

- **If the IR Center Sensor reading is less than 10, the robot stays still.**
- **If the IR Center Sensor reading is between 10 and 200, the robot spins right with its speed equal to 4 times the current IR Center Sensor value and it plays the Musical Instrument 9 for 0.5 beat.**
- **If the IR Center Sensor reading is between 200 and 450, the robot spins left with its speed equal to 4 times the current IR Center Sensor value and it plays the Musical Instrument 16 for 0.5 beat.**
- **If the IR Center Sensor reading is larger than 450, the robot rolls forward at Speed=1023 to escape from the object and it plays the Musical Instrument 21 for 0.5 beat.**

This Condition-Action list indicates that we are dealing with a situation whereas one and only one condition can happen at any one time during the runtime performance of this robot, thus we will need to use an IF-ELSE-IF structure for our solution to this problem.

The program "IR-based-Maneuvers.sb2" describes the solution approach used:

1) The Initial Event is solved by playing Melody 2 out of the CM-150's speaker and Variable Speed is initialized to zero (see Fig. 4.20).

Fig. 4.20 Initial Event for program "IR-based-Maneuvers.sb2".

2) The main algorithm (Fig. 4.21) uses an outer FOREVER loop to represent the World-Robot Interactions Cycle (previously depicted in Fig. 1.12). This loop contains an IF-ELSE-IF structure representing the Condition-Action/Input-Output listing as detailed out previously. During each loop, the PC would evaluate/execute this IF-ELSE-IF structure in the following manner:

- It acquires the latest IR Center Sensor reading and saves it in Variable IR_Sensor.
- Next, it enters the IF-ELSE-IF structure and checks the first IF statement, i.e. the condition (IR_Sensor < 10). If this condition is TRUE, it turns off both motors, it then exits the whole IF-ELSE-IF structure and performs another iteration of the FOREVER loop. But if this condition is FALSE, i.e. (IR_Sensor >= 10), the PC goes on to check the next second IF statement which is based on the condition (IR_Sensor < 200).

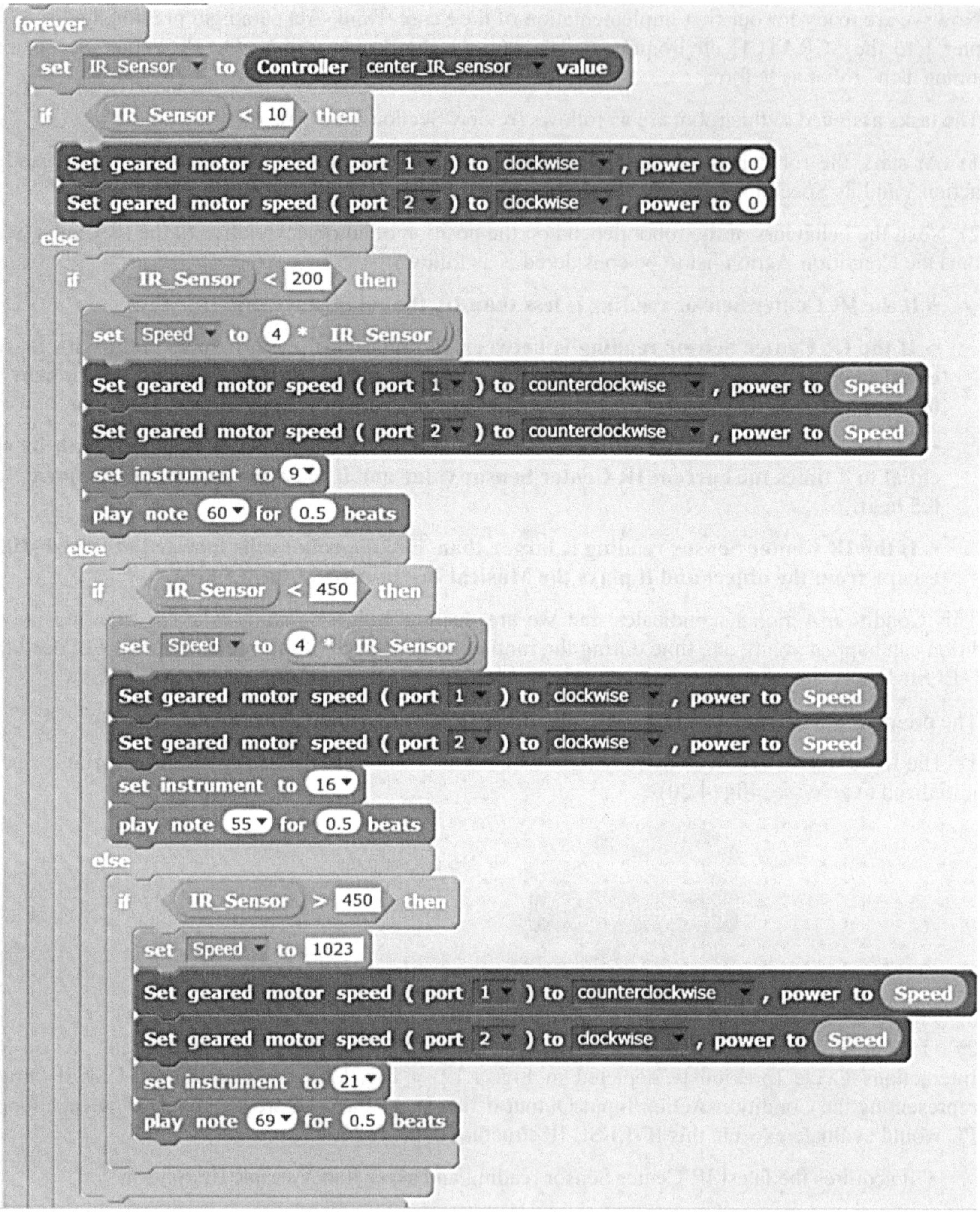

Fig. 4.21 Main Algorithm used in program "IR-based-Maneuvers.sb2".

- Please note that according to the considered condition-action listing, the PC is supposed to check if IR_Sensor is between 10 and 200, but here it uses a condition that involves only checking for (IR_Sensor < 200). Is that sufficient to do what is required? And the answer is Yes and that it is

OK to just check for the condition (IR_Sensor < 200) here, because the condition (IR_Sensor >= 10) is implicitly TRUE when the condition (IR_Sensor < 10) evaluates to FALSE (see first IF).

- Thus, if the second condition (IR_Sensor < 200) turns out to be TRUE (meaning that IR_Sensor is between 10 and 200), the PC assigns the value of "4*IR_Sensor" to Speed and makes the robot spin right at this value of Speed. The PC also plays Note 60 on Instrument 9 for 0.5 beats, then it exits this IF-ELSE-IF structure and makes another iteration of the FOREVER loop. However, if the condition (IR_Sensor < 200) turns out to be FALSE, i.e. (IR_Sensor >= 200), then the PC goes on to check the third IF statement.

- Similarly, with the third IF, the PC is checking to see if IR_Sensor is between 200 and 450. If this is TRUE, then the robot spins left with a Speed equal to "4*IR_Sensor", plays Note 55 on Instrument 16 for 0.5 beats, exits this IF-ELSE-IF structure and continues with a new iteration of the FOREVER loop. If the condition of the third IF turns out to be FALSE (i.e. (IR_Sensor >= 450), then the fourth IF is executed next.

- If the fourth IF, i.e. (IR_Sensor > 450), turns out to be TRUE, then the PC would run forward at Speed=1023 and plays Note 69 on Instrument 21 for 0.5 beats. This maneuver is designed to make the robot run away from under the approaching object. Once away from the object's influence, the IR_Center Sensor would read zero, and the robot would stop and be ready to "play again".

Video 4.10 gives a verbal explanation of the program "IR-based-Maneuvers.sb2" and shows its runtime performance.

Fig. 4.22 shows the program "IR-based-Maneuvers-Say.sb2" which is based on the previous program "IR-based-Maneuvers.sb2", with appropriate "Text Balloons" generated with SAY blocks from the LOOKS menu in the SCRIPTS tab. Video 4.11 shows the runtime performance of this program.

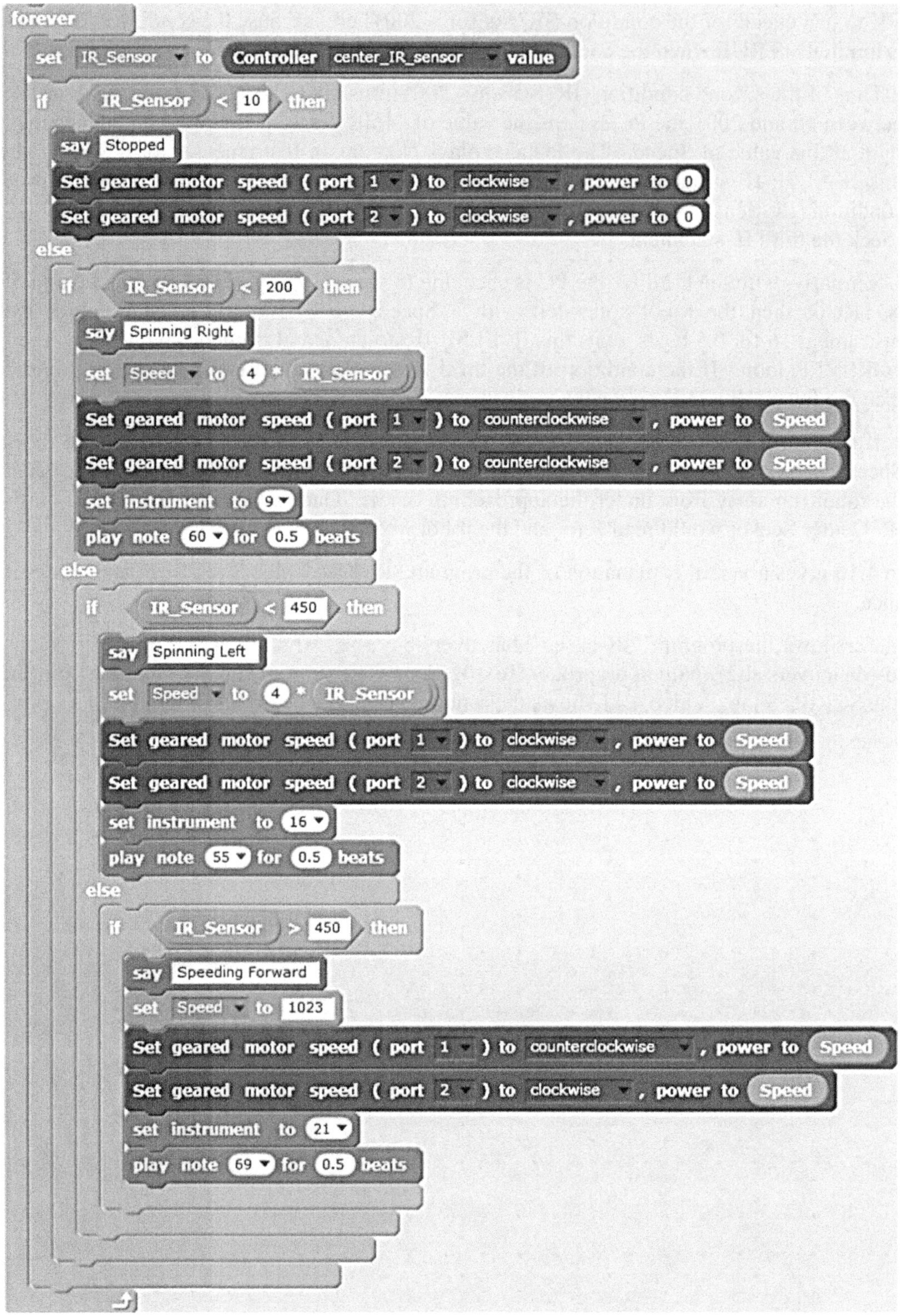

Fig. 4.22 Main Algorithm used in program “IR-based-Maneuvers-Say.sb2”.

Included in the example codes is the "IR-based-Maneuvers-BW.sb2" program which illustrates the use of the "BROADCAST & WAIT" technique to create program behaviors similarly to using Functions in TASK codes. Please see the Review Questions Section of this Chapter for more details.

4.5 "Avoider" Maneuvers using Left and Right IR Sensors

In this section, the previous "Spinning Top" platform is still used, but it is renamed "Avoider" to emphasize the types of maneuvers described herein. The "Avoider" will be using its Left and Right IR Sensors to detect a single object and to accomplish tasks described in the following Condition-Action list:

1) If there is no obstacle in front of the Avoider, it can go forward.

2) If there is an object on the Left of the Avoider, it should back away and turn to the Right to avoid the obstacle.

3) If there is an object on the Right of the Avoider, it should back away and turn to the Left to avoid the obstacle.

4) If there is an object in Front of the Avoider, e.g. a Corner Wall, it should first go Straight Back to avoid the obstacle then turn Left and Backward to steer away from the obstacle.

This is another familiar programming situation of **choosing one and only one alternative among several possible alternatives**, thus an **IF-ELSE-IF** structure should do an adequate job when combined with an **Outer FOREVER Loop**. The SCRATCH solution is named "Avoider-2.sb2". It starts with the Playing of Melody 2 and the SETting of Variable Speed to 256. The main algorithm is described in Fig. 4.23:

1) The FOREVER Loop starts out by reading in the latest values from the IR Left and Right Sensors and saves them respectively into Variables IR_Left and IR_Right.

2) Next is an IF-ELSE-IF structure with 4 mutually exclusive cases:

- For the First IF block, if the values of "IR_Left" AND "IR_Right" are both less than 10 (i.e. no obstacle in front of Avoider) then a "Go Forward" Text Balloon pops up in the Stage area and the robot goes forward. The PC then exits the IF-ELSE-IF structure and begins another iteration of the FOREVER loop. However, if this condition evaluates to FALSE, meaning that either "IR_Left" or "IR_Right" have values larger than 10, or in other words when **either** IR sensor sees an object, the PC would need to check further by executing the Second IF block.

- In the Second IF block, the condition ((IR_Left > 100) AND (IR_Right < 10)) represents a situation where there is an object on the Left of the robot and none on its Right side. If this condition evaluates to TRUE, a "Right & Back" Text Balloon appears in the Stage and the robot turns Right by Backing away to avoid the obstacle on its left. Next, the PC exits the IF-ELSE-IF structure and starts the next iteration of the FOREVER loop. If this condition evaluates to FALSE, then the Third IF block is executed.

- Similarly, the condition ((IR_Right > 100) AND (IR_Left < 10)) used by the Third IF block represents a situation where there is an object on the Right of the robot and none on its Left. If this condition evaluates to TRUE, a "Left & Back" Text Balloon appears in the Stage and the robot Backs away and turns Left to avoid the obstacle on its Right. Next, the PC exits this IF-ELSE-IF structure and starts a new iteration of the FOREVER loop. If this condition evaluates to FALSE, then the fourth IF block is examined next.

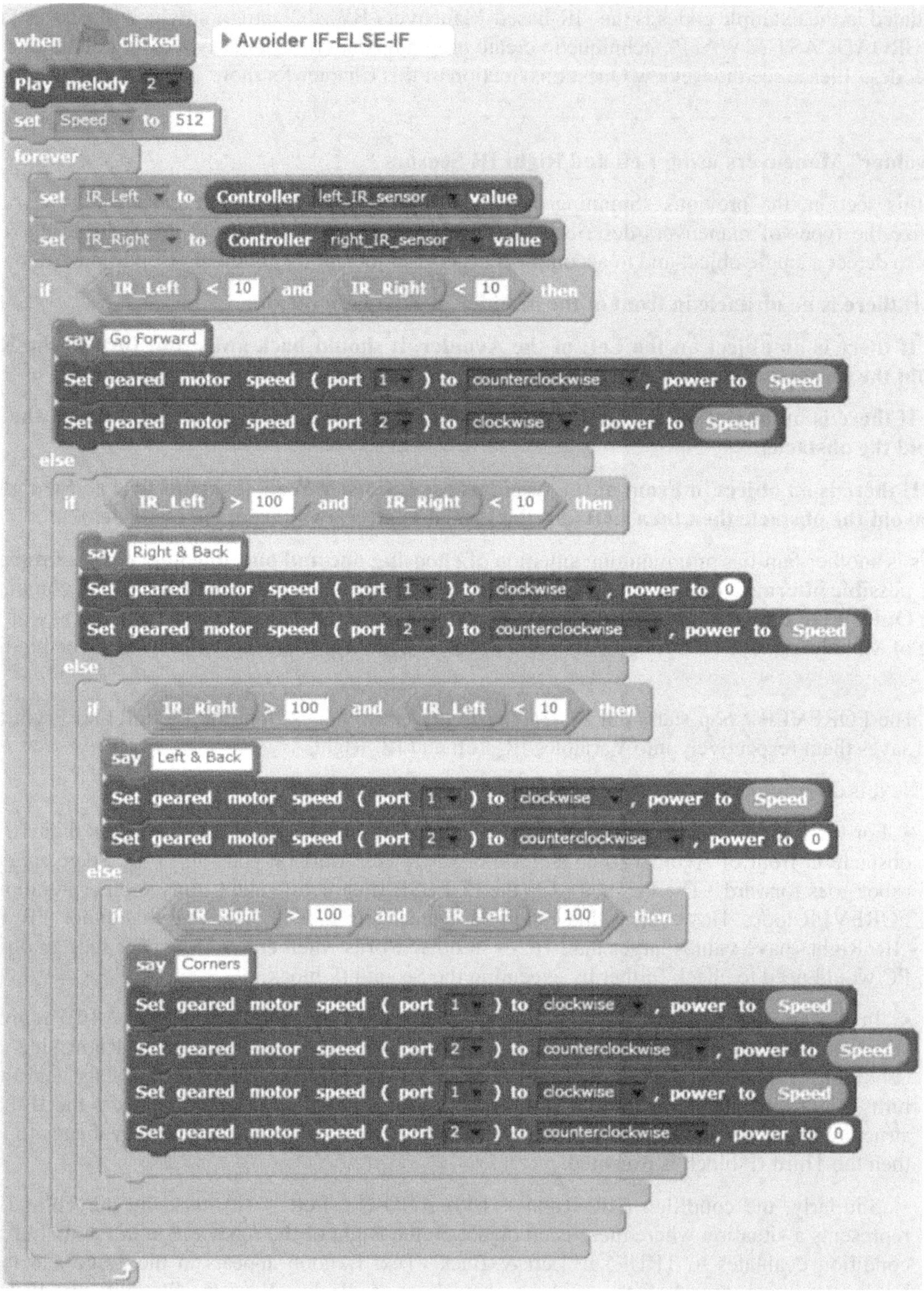

Fig. 4.23 Main SCRIPT used in program “Avoider-2.sb2” (IT-ELSE-IF used).

• The Fourth IF block uses this condition ((IR_Left > 100) AND (IR_Right > 100)) to represent a situation when there is an object close enough or large enough in front of the robot such that both IR Sensors report data values larger than 100. If this condition is TRUE, the "Corners" Text Balloon appears on the Stage and the robot would first move straight back then make a "left-back" turn to steer it away from any potential corner or even a cul-de-sac (the reader is encouraged to try out a "right-back" turn instead and see the subsequent robot behavior).

Video 4.12 shows that this Avoider-2 algorithm has a good performance, but is less responsive than its TASK counterpart ("Avoider-2.tskx" and Video 3.13).

The next SCRATCH project "Avoider-3.sb2" offers a different solution technique to these maneuvering tasks. **This new approach still relies on the same Condition/Action listing as described at the beginning of this section 4.5, but it distills different behavior patterns from this list:**

1) We must identify a "default" action which in this case is "Going Forward".

2) We will be using several simple IFs in parallel instead of a single big IF-ELSE-IF structure like before. **We also need to do a deeper analysis of the condition/action list to retain only the "essential" condition/action pairs.**

3) We also **need to apply a Delay Time to each action made** (this part was not needed when the IF-ELSE-IF structure was used). This Delay Time can be tuned for optimal performance.

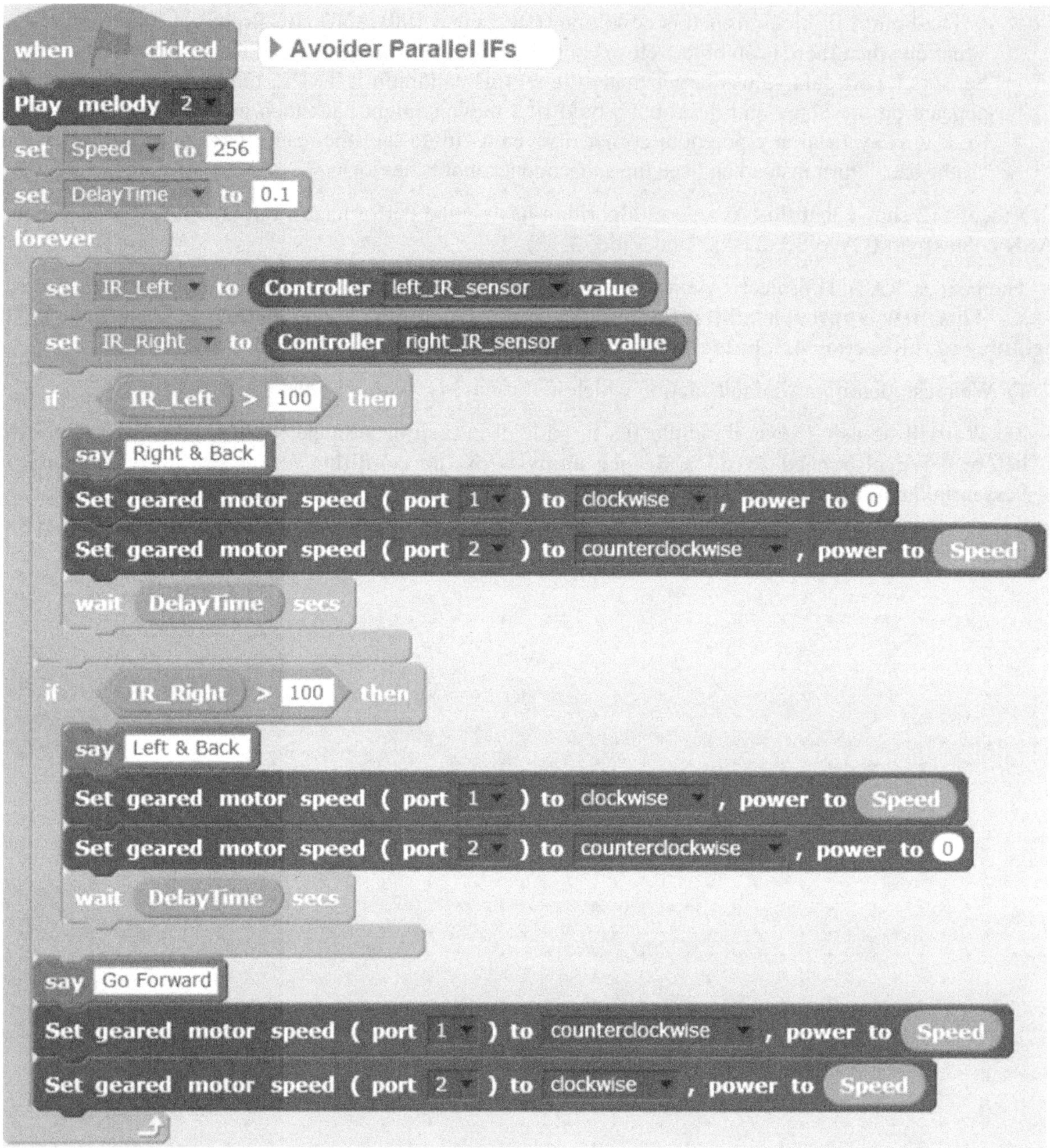

Fig. 4.24 Main SCRIPT used in program "Avoider-3.sb2" (parallel IFs used).

Fig. 4.24 shows the revised FOREVER loop:

1) At the start of the FOREVER LOOP, Variables IR_Left and IR_Right are updated with the latest data input from the IR Left and Right Sensors. Next are two parallel IF blocks:

- The first IF block checks only on the condition for an object on the robot's Left (conditional expression "IR_Left > 100"). If this is TRUE, the robot backs away and turns right to steer itself away from the obstacle. A short delay (0.1 second) is necessary for the robot to stay in this "action" to make this algorithm work properly. The reader can try different values for the time delay to find the optimal robot performance.

- The second IF block represents the condition when there is an object on the robot's right, then the robot backs away and turns left to steer itself clear from the obstacle. A 0.1 second delay (user-tunable) is also necessary for the robot to stay in this "action" to make this algorithm work properly.

2) **Please note that the "default" maneuver is always programmed as the "Last Action" ("Going Forward" in this case).**

The reader can readily see that this "Avoider3.sb2" version is more compact than the "Avoider-2.sb2" version. Video 4.13 shows the runtime performance of this algorithm.

The next project "Avoider-4.sb2" offers a third approach to the same condition/action list shown at the beginning of Section 4.5, but now using "REPEAT UNTIL" loops (see Fig. 4.25). REPEAT UNTIL Loops are available in the CONTROL menu of the SCRIPT tab.

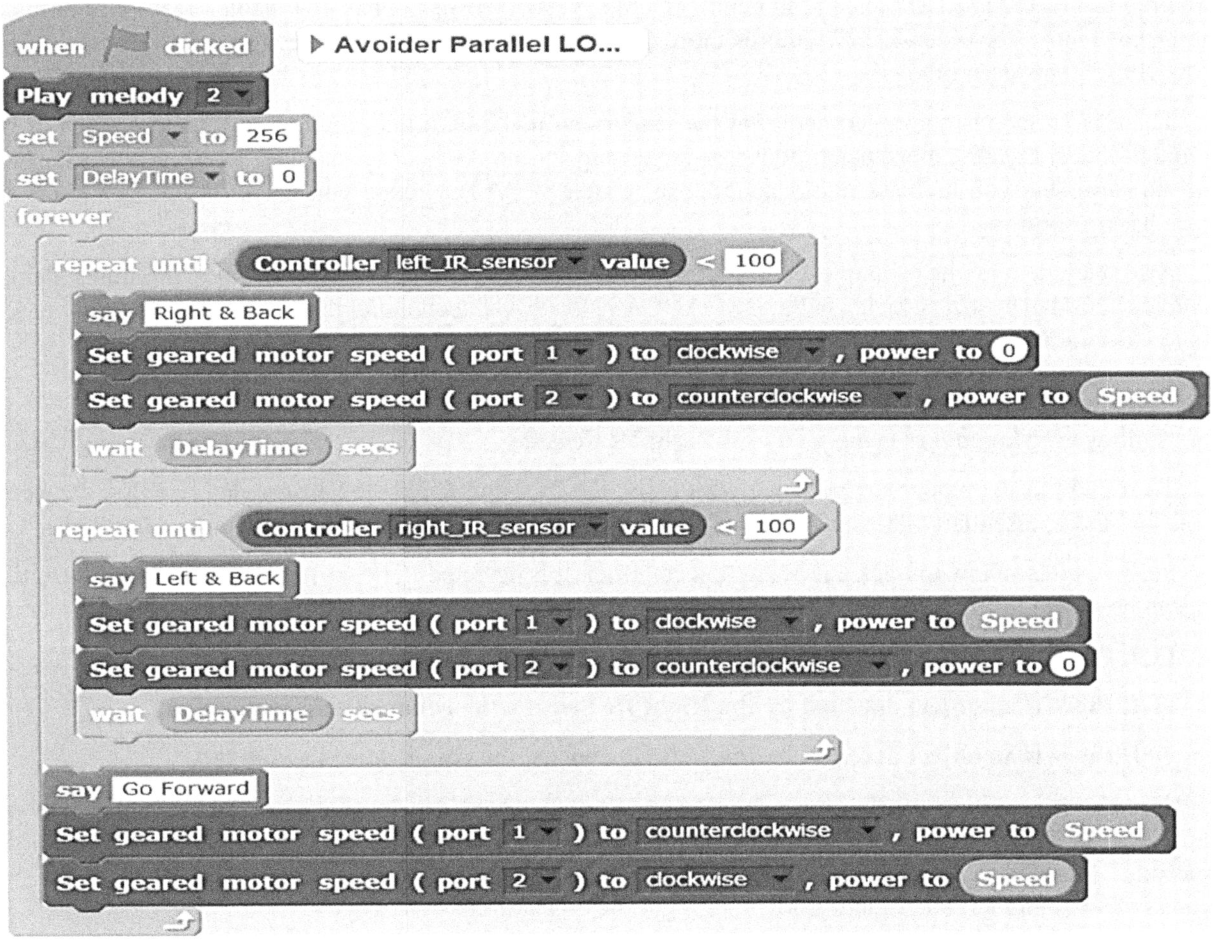

Fig. 4.25 Main SCRIPT used in program "Avoider-4.sb2" (parallel REPEAT UNTIL loops).

Fig. 4.25 shows that it also uses the "Default Maneuver" approach (same as "Avoider-3.sb2"), and the parallel REPEAT UNTIL loops match up with the parallel IFs (location wise). But there are three differences:

1) Delay Times are not needed for REPEAT UNTIL loops (because they are set to zero), but the reader is encouraged to try different values for Parameter DelayTime to see how they affect the overall robot behavior.

2) The user needs to make sure to read in the most recent data values for all Variables set up in the Conditional Expressions used by the REPEAT UNTIL loops. Thus direct "Controller Values" are used on each Loop's conditional expression instead of the Parameters "IR_Left" or "IR_Right" as in "Avoider-3.sb2".

3) The user also needs to note that the Conditional Expression used in a REPEAT UNTIL loop is the Negation of an Equivalent Conditional Expression used in an IF block (please review and contrast Figs. 4.24 and 4.25).

Video 4.14 shows the runtime performance of the program "Avoider-4.sb2".

Does the reader have a preference over one of these 3 versions of Avoider? The big IF-ELSE-IF structure of "Avoider-2.sb2" may be too cumbersome. "Avoider-3.sb2" has a "tunable" feature in terms of the Delay Time. "Avoider-4.sb2" requires more executions of the "Read" type Extension Command which costs 100-150 ms each time.

There is an interesting experiment that the reader can make with the previous different versions of the Avoider codes: create a maze using books or boxes and let the Avoider robot roam freely through it. You will be surprised with its behaviors depending on which SCRATCH code you used, especially when your maze has a cul-de-sac.

Also included in the example codes are alternate versions of the above Avoider programs named "Avoider-2/3/4-BW.sb2" which use the BROADCAST & WAIT technique, the reader is encouraged to test out those solutions on his or her own.

4.6 "Follower" Maneuvers using Left and Right IR Sensors

To round out the reader's experience with using the Left and Right IR Sensors, let's apply a **"reverse logic"** on them, essentially let's make the robot follow an object when it detects something in front of it.

Let's first learn how to track a moving object - when it is detected. The corresponding Condition/Action list can be as follows:

1) **If there is no object detected, the robot should stay still (default behavior).**

2) **If there is an object detected by the Right IR Sensor, the robot should swing right.**

3) **If there is an object detected by the Left IR Sensor, the robot should swing left.**

Should we also plan for the case when an object is detected and located right in front of the robot? Should it "stop" or should it show that it already "locks in" on the object in some other way? From the preceding Avoider SCRATCH coding work, we should be able to let the robot bounce between its two actions of swinging left and right to show that it got the object in its sight, right?

The program "Follower-1.sb2" was derived from the previous discussions using REPEAT UNTIL Loops (i.e. "Avoider-4.sb2") within a Main FOREVER Loop. The algorithm used is shown in Fig. 4.26:

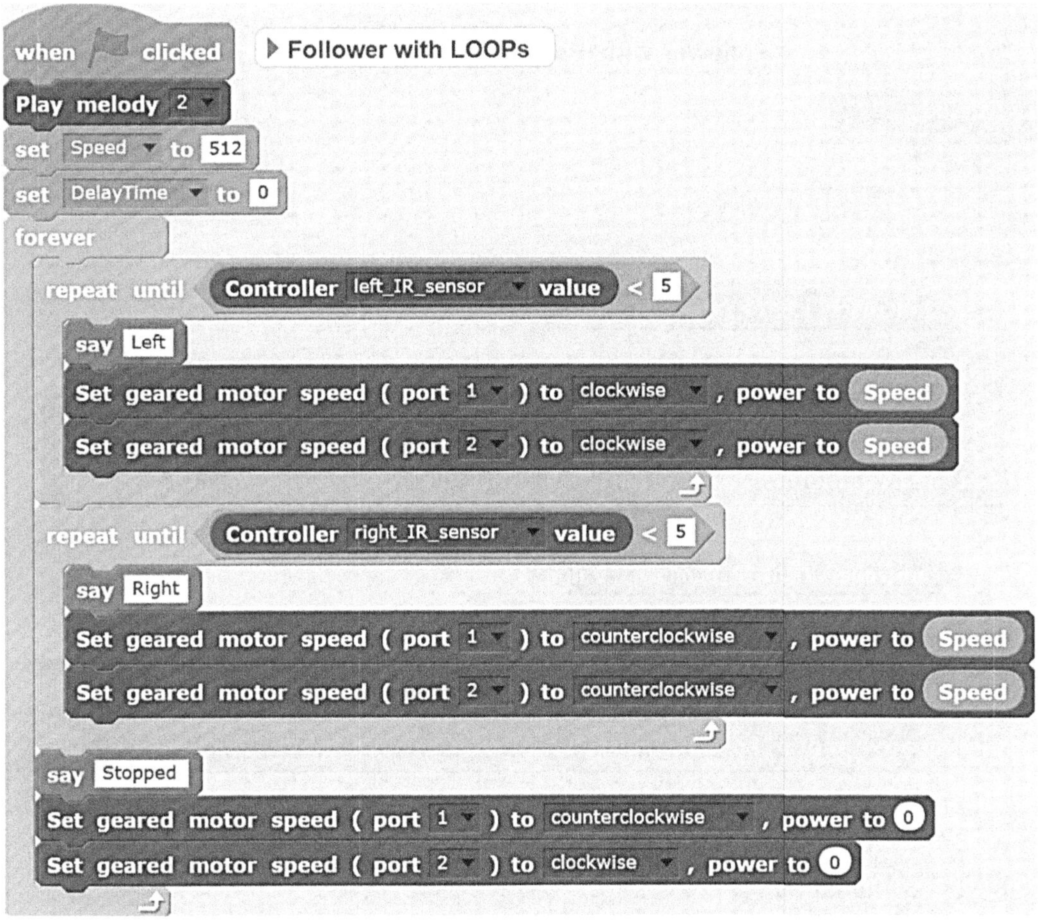

Fig. 4.26 Main SCRIPT used in program "Follower-1.sb2" (parallel REPEAT UNTIL loops).

1) Essentially, whenever any IR Sensor (Left or Right) picks up on an object, the robot rotates respectively towards the same side where that IR Sensor is located. These actions are implemented by the two REPEAT UNTIL Loops. Please note that a low detection threshold value of "5" is used so that the robot can detect an object as "far" as possible!

2) The default "Action/Maneuver" is to stop and it is used at the bottom of the FOREVER Loop.

Also, please note the structural similarity between "Avoider-4.sb2" and "Follower-1.sb2", except for the actual Conditional Expressions and Actions used, of course.

Video 4.15 shows the runtime performance of this algorithm. **Please note the effect of the object's size on how the robot would tract the object.**

Fig 4.27 shows another algorithm using parallel IFs instead, but yielding the same object tracking result but with a staccato rhythm. Please note that a small Delay Time is no longer needed as compared to an equivalent TASK code, as the SCRATCH/R+SCRATCH tool chain has plenty of system delay times to contend with already. Please also note the "reverse logic" in the Conditional Expressions used between "Follower-1.sb2" and "Follower-1-IFs.sb2" (i.e. REPEAT UNTIL vs. IF). See Video 4.16 for a runtime performance of the program "Follower-1-IFs.sb2".

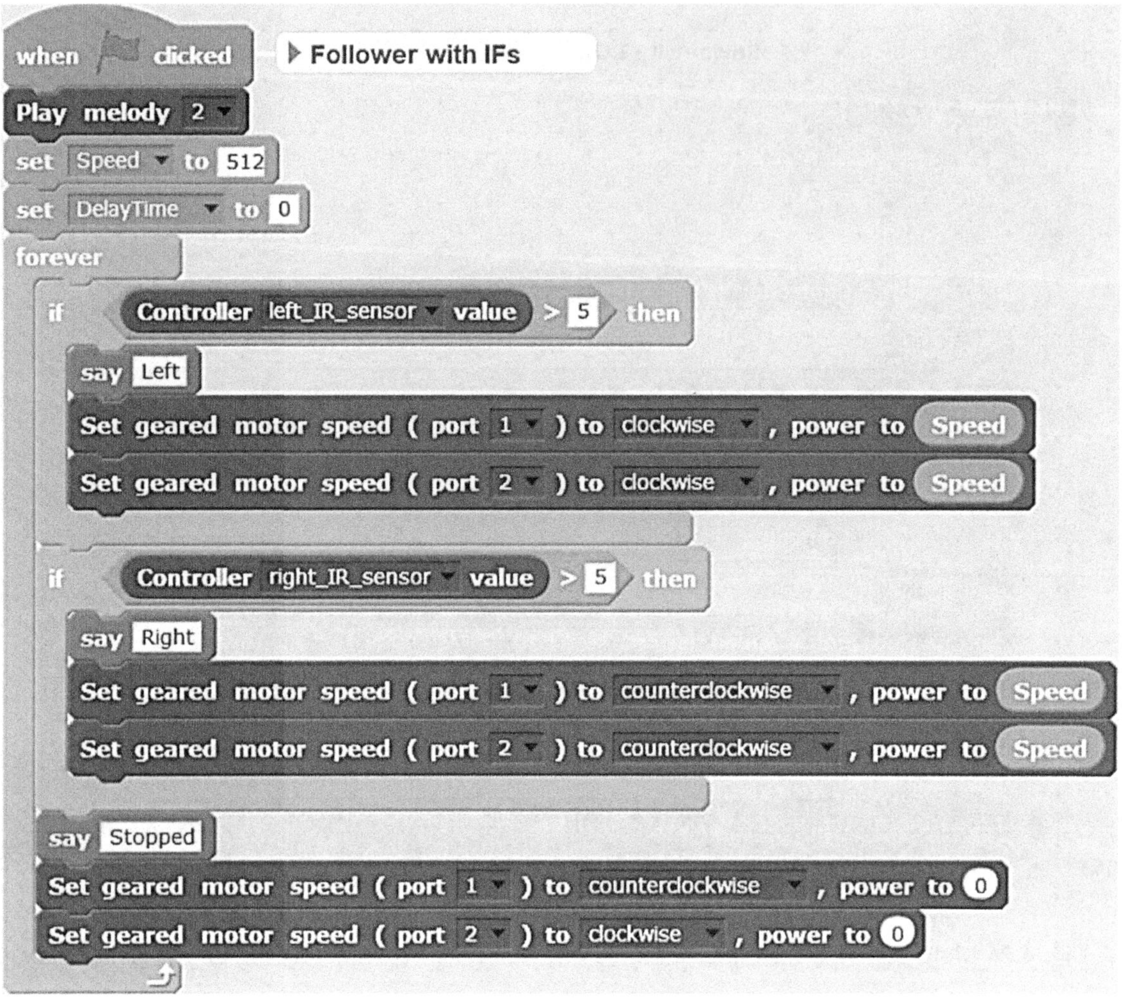

Fig. 4.27 Main SCRIPT used in program "Follower-1-IFs.sb2" (parallel IF blocks).

In the next "Follower" applications, a new Condition/Action pair is added to the previous list:

When the object gets a bit closer to the robot (i.e. threshold value of 10 applied to both IR Sensors), the robot should get even closer to the object by going Forward. At this point during runtime, if the user can move the object away from the robot at a proper speed, ones would see the effect of the robot "following" the object from a short distance away.

There are two possible solutions: one using a **combination of REPEAT UNTIL and IF-ELSE-IF** structures (i.e. Follower-2.sb2), and another one using **only an IF-ELSE-IF** structure (i.e. Follower-3.sb2).

The Main algorithm of the program "Follower-2.sb2" (see Fig. 4.28) uses a default maneuver (i.e. "Stop" the robot) and an outer REPEAT UNTIL Loop which encompasses an inner IF-ELSE-IF structure that determines the unique and appropriate action among 3 possible and mutually exclusive alternatives: "Left", "Right" or "Forward" maneuvers:

1) When no obstacle is detected (i.e. (IR_Left < 5 AND IR_Right < 5) is TRUE), the REPEAT UNTIL Loop is skipped and program flow control is transferred to the bottom of the FOREVER LOOP, i.e. saying "Stopped" and turning power off to Ports 1 and 2 motors. Thus, the robot would stay still at its default behavior.

2) When <u>either</u> IR Sensor detects an object (i.e. (IR_Left < 5 AND IR_Right < 5) evaluates FALSE), the REPEAT UNTIL Loop is entered:

- Next, the PC checks further with an IF-ELSE-IF structure to see if the IR Left Sensor is the only one detecting the object (e.g. threshold "5" exceeded in the First IF block) then the robot says "Left" and goes Left.

- But if this first IF condition turns out to be FALSE, the PC next checks the Second IF block to see if the IR Right Sensor is the only one detecting the object (e.g. threshold "5" exceeded in Second IF), then the robot says and goes "Right".

- However, if this Second IF Condition evaluates to be FALSE also, then the PC next checks to see if the object is close to both IR Sensors. If the conditional expression for the Third IF block turns out to be TRUE (i.e. both IR Sensors exceed a threshold of "75"), then the robot says and goes "Forward".

- But if this Third IF condition turns out to be FALSE, the PC reaches the end of the IF-ELSE-IF structure without determining a concrete action, so the PC then loops back to check on the condition of the REPEAT UNTIL Loop, i.e. ((IR_Left < 5) AND (IR_Right < 5)). If this condition is still FALSE (i.e. either IR Sensor sees an object), it then searches through the IF-ELSE-IF structure again. If this condition becomes TRUE (i.e. no object found), the PC finally quits the REPEAT UNTIL Loop. It next SAYs "Stopped" and stops the robot and then repeats the outermost FOREVER LOOP all over again.

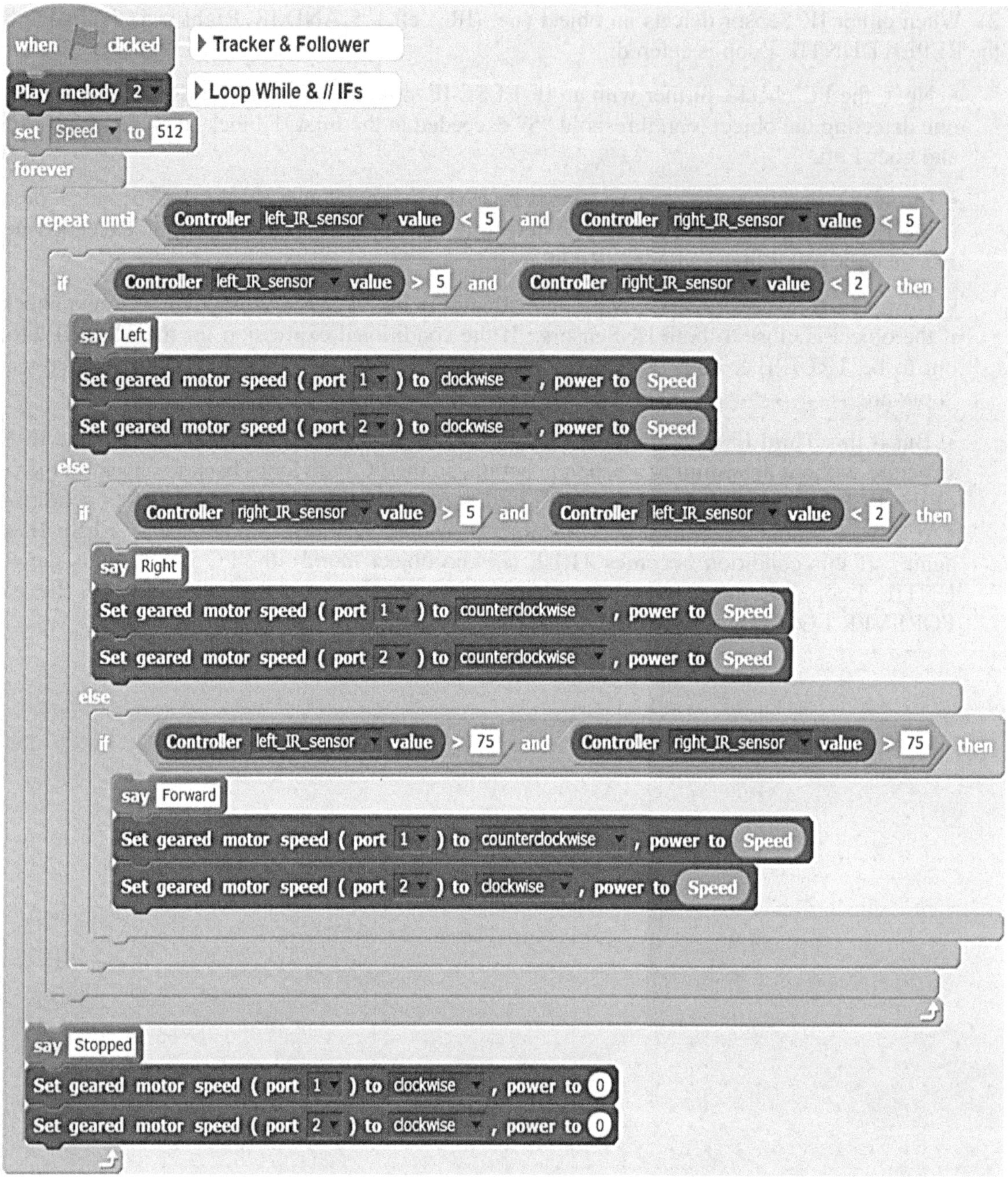

Fig. 4.28 Main SCRIPT used in program "Follower-2.sb2"

(Combination of LOOP and IF blocks).

Please note, once again, that when using REPEAT UNTIL Loops, we need to refresh and use the most recent IR Sensor Data inputs in all our conditional statements, just before entering the REPEAT UNTIL Loop and as the last actions of the REPEAT UNTIL Loop (i.e. do not use a SCRATCH parameter that had previously been set to an IR Sensor Data block – please review Fig. 3.18 and its explanation). Please see Video 4.17 for the runtime performance of "Follower-2.sb2".

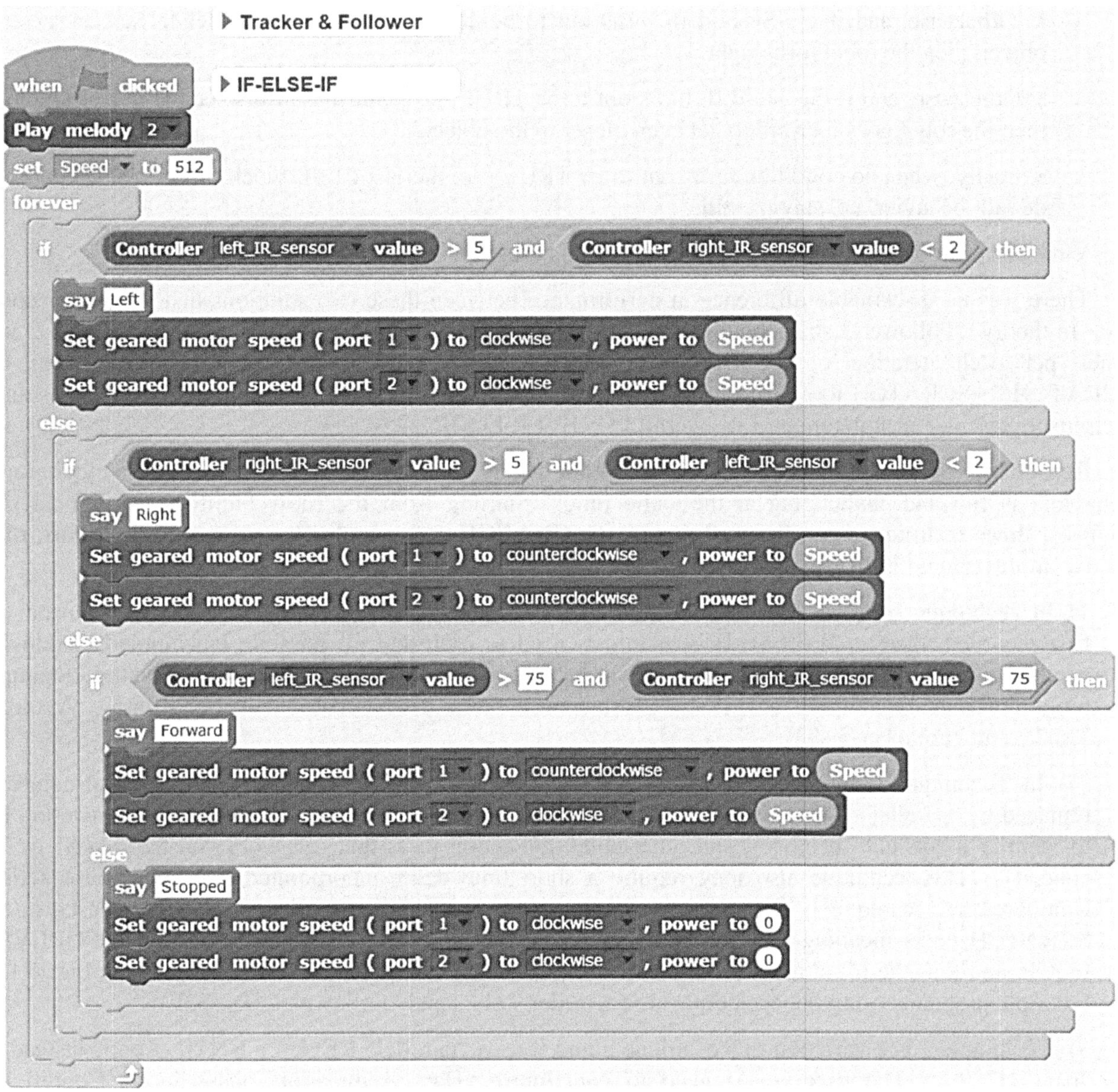

Fig. 4.29 Main SCRIPT used in program "Follower-3.sb2"

The algorithm used in the next program "Follower-3.sb2" is described in Fig 4.29. It is more straight forward so it is easier to explain:

1) It uses an outermost FOREVER LOOP (as always) and the latest IR Sensor values are used in the Conditional Expressions used by the IF-ELSE-IF structure (i.e. "Controller Value" blocks) to ensure as much "real-time" operations as possible.

2) The familiar IF-ELSE-IF structure now represents 4 mutually exclusive condition/action pairs for "Left", "Right", "Forward" or "Stop" (please note that the IF-ELSE-IF structure in the previous "Follower-2.sb2" handles only 3 condition/action pairs):

- If the First IF is TRUE, i.e. only the IR Left Sensor sees the object, then the robot turns left.

• Otherwise, and if the Second IF turns out to be TRUE, i.e. only the IR Right Sensor sees the object, then the robot turns right.

• Otherwise, and if the Third IF turns out to be TRUE, i.e. both IR sensors see the object up close, then the robot goes forward to get even closer to the object.

• Lastly, when no condition turns out to be TRUE (i.e. the last ELSE block), the robot executes its default behavior, i.e. staying still.

Video 4.18 shows the run-time performance of this program "Follower-3.sb2".

There was no discernable difference in performance between these two solutions that the author could see. In theory, "Follower-3.sb2" should be "slightly faster" because it accesses the sensor ports from 2 to 6 times per each iteration of the overall FOREVER LOOP, i.e. it uses less resources from the SCRATCH/R+SCRATCH tool chain as compared to "Follower-2.sb2" which uses from 2 to 8 accesses to the sensor ports per each iteration of its overall FOREVER LOOP.

Recapping Sections 4.4, 4.5 and 4.6, the author's goal is to show that programming "autonomous" behaviors is fun and challenging at the same time! Starting from the (now familiar) condition/action analysis, **three techniques can be used to create SCRATCH codes and after some initial actions, they all use an outermost FOREVER LOOP for the main algorithm:**

1) In Technique 1, for each iteration of the FOREVER LOOP, latest readings from the used sensors are first recorded, then an IF-ELSE-IF structure is used to represent all possible but mutually exclusive condition/action pairs that correspond to robot behaviors required to fulfill a given challenge. Example programs using Technique 1 are "IR-based-Maneuvers.sb2", "IR-based-Maneuvers-Say.sb2", "Avoider-2.sb2", and "Follower-3.sb2".

2) In Technique 2, the sensors reading task is still required first, but the IF-ELSE-IF structure is replaced by "parallel" IFs. This technique requires a deeper analysis of the original condition/action list to identify a "default" behavior, and other condition/action pairs that can work "on their own" or "in concert". This technique also may require a short time delay incorporated into each robot action. Compared to Technique 1, Technique 2 usually results in a smaller number of statements used in the SCRATCH code, meaning that the PC should take less time to do each iteration of the FOREVER LOOP, and thus the robot may become more responsive to changes in the "World" (review Fig. 1.12). Example programs using this technique are "Avoider-3.sb2" and "Follower-1-IFs.sb2".

3) Technique 3 is a variation of Technique 2 and it uses "parallel" REPEAT UNTIL Loops instead of "parallel" IFs. The user would need to **coordinate** "Data Acquisition" tasks in its Conditional Statements and at appropriate parts of the REPEAT UNTIL Loops. Thus, the user should expect more system delays as compared to the other two Techniques, as "Data Acquisition" commands cost 70-100 ms each to go through the SCRATCH/R+SCRATCH/CM-50 hardware/software chain. Example programs using Technique 3 are "Avoider-4.sb2" and "Follower-1.sb2". "Follower-2.sb2" is a hybrid solution combining Techniques 3 and 1.

4.7 Smart Avoider

This section mirrors Section 3.7.4 and shows how to combine Autonomous-Behavior and Remote-Control techniques so that the Avoider robot (Fig. 3.10) can "save itself" when, on purpose, the user drives the robot straight into an obstacle!

The SCRATCH solution "SmartAvoider.sb2" is adapted from the TASK code "SmartAvoider.tskx" and SCRATCH code "RC-FlashDancer-bw.sb2":

1) An “Event Programming” approach is used whereas several SUB-SCRIPTs are created to handle various key-based and message-based events. The two “key-based” events are shown in Fig. 4.30 and are designed to set Parameters “Speed”, “R_Threshold” and “L_Threshold” with their appropriate values depending on whether Key 1 (for low speed) or Key 3 (for high speed) is pressed at run-time. A higher speed setting would require a lower threshold value because the Avoider would have initiate evasion maneuvers sooner to avoid the obstacle. The Forever loop in Fig. 4.30 is used to display the “live” IR Sensors values when the Avoider is manually placed at the appropriate “avoidance” distance with respect to the obstacle.

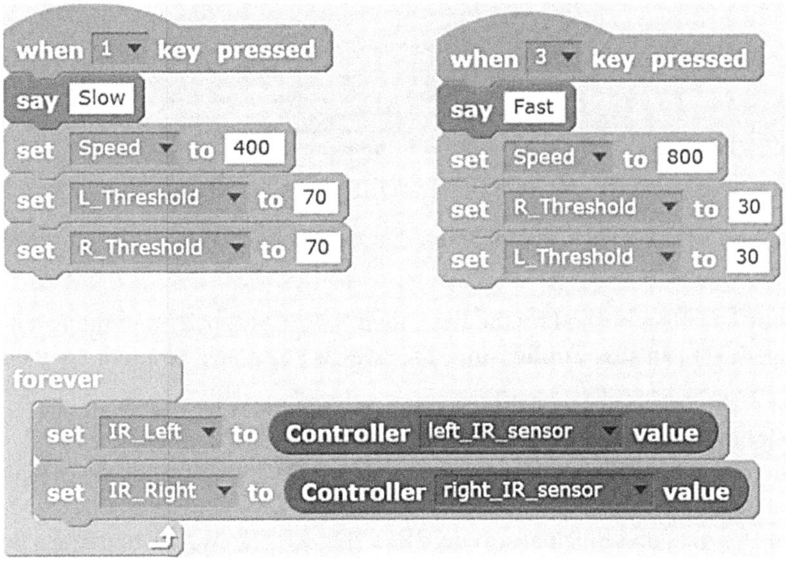

Fig. 4.30 SUB-SCRIPTs for “1 or 3” Key-based Event and Data Display Utility for setting up NIR Thresholds in “SmartAvoider.sb2”.

All the robot maneuvers are triggered by “message-based” events: “Forward”, “Backward”, “Right”, “Left” and “Stop”. Fig. 4.31 shows the SUB-SCRIPTs used for “Forward” and “Backward” events.

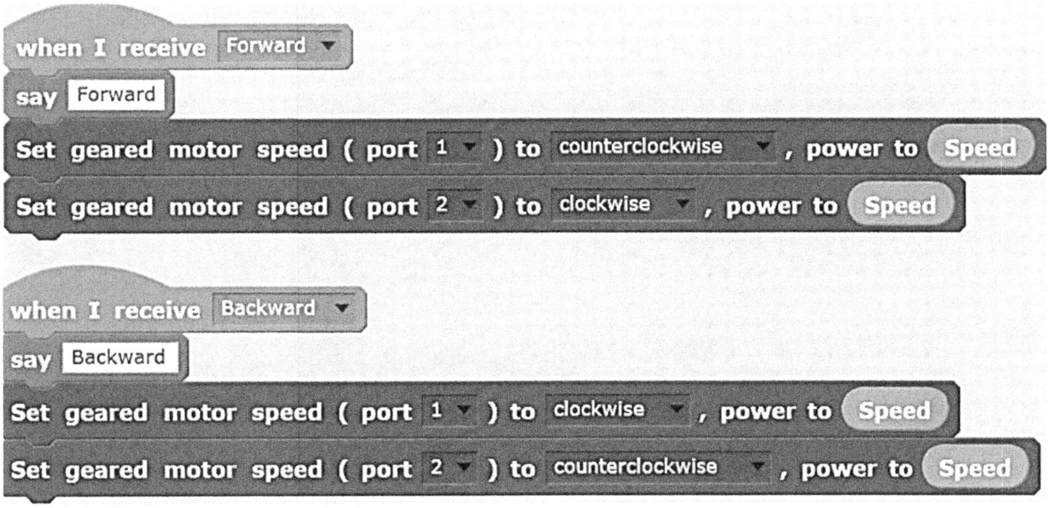

Fig. 4.31 SUB-SCRIPTs for “Forward” and ”Backward” maneuvers in “SmartAvoider.sb2”.

2) The MAIN SCRIPT get initiated with PLAYing Melody 3 and initializing Parameters Speed, L_Threshold and R_Threshold with their respective values (see Fig. 4.32).

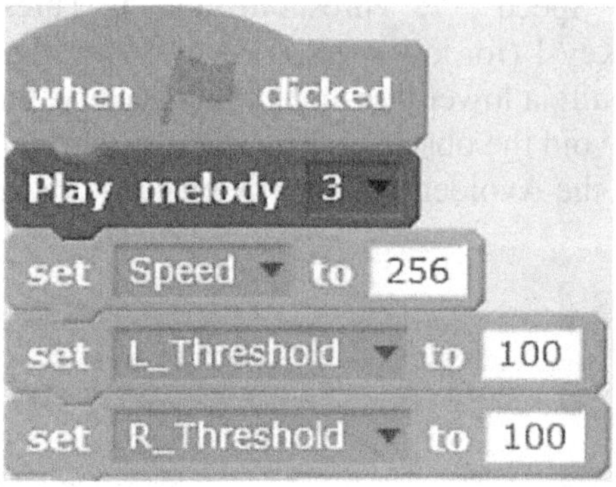

Fig. 4.32 MAIN SCRIPT Initializations in “SmartAvoider.sb2”.

3) Next, the MAIN FOREVER Loop is entered, a nested IF-ELSE-IF structure is used to sort out 3 main situations:

a) If NO key has been pushed by the user, then SAY “Stopped” and the robot’s motors are turned off (see Fig. 4.33). **In this application, the user is supposed to push only one key at any time.**

b) ELSE, i.e. the user has pressed on some key, then the robot needs to check on the values of its Left and Right IR Sensors against their respective Threshold values (see Fig. 4.33). If these Sensor readings are below the Thresholds, the robot is still far enough from the obstacle, therefore it is appropriate to start processing the actual value of the key being pressed. Next, the arrow keys are checked one by one, and the appropriate BROADCAST & WAIT messages are sent to the appropriate Event-Processing SUB-SCRIPTs like the ones shown in Fig. 4.31.

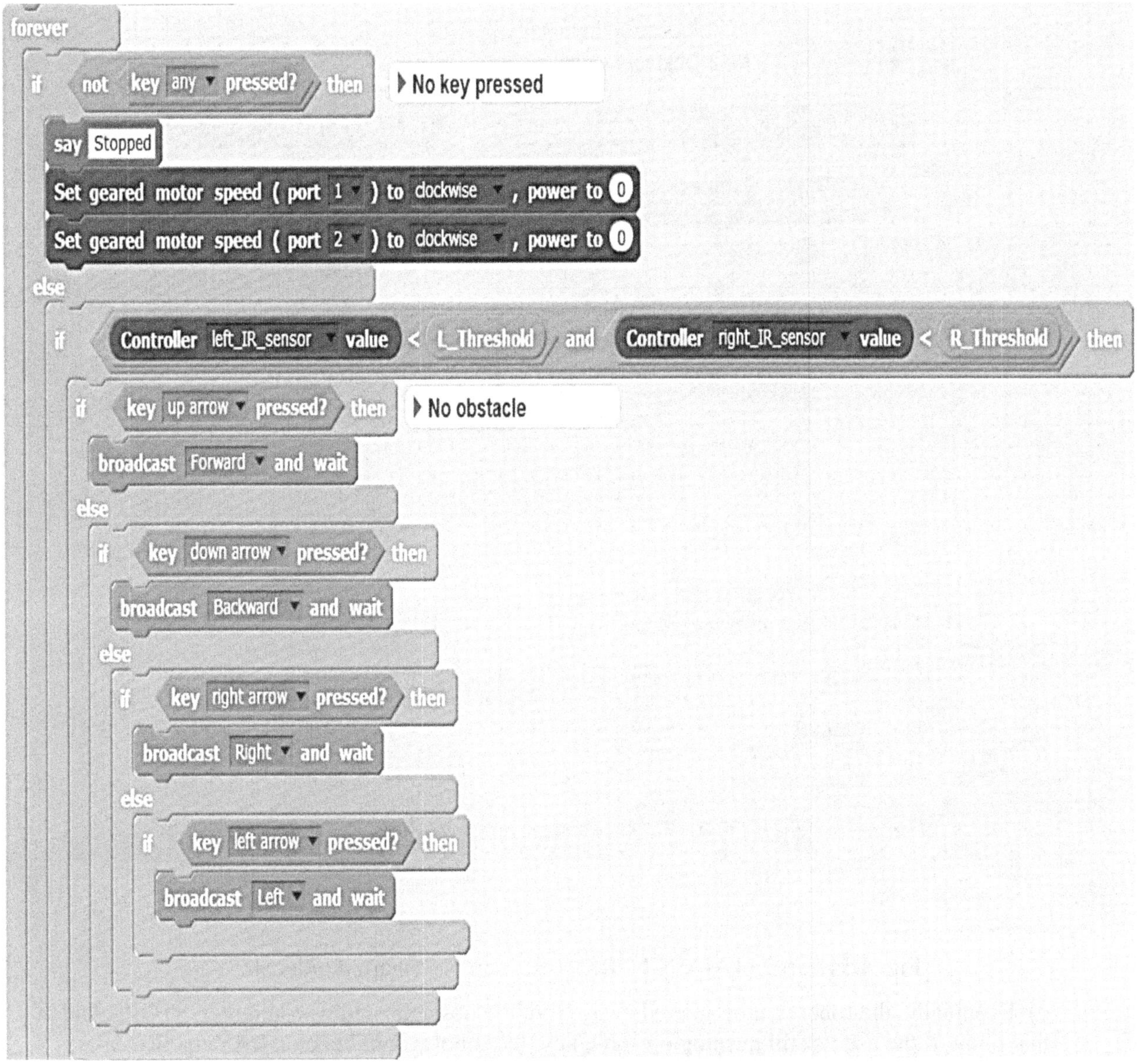

Fig. 4.33 Part 1 of MAIN FOREVER Loop in "SmartAvoider.sb2".

c) However, as the user navigates the robot closer to the obstacle, one of the IR Sensors is bound to get a reading higher than its corresponding threshold value, triggering the top ELSE as shown in Fig. 4.34, and the robot first needs to stop (i.e. BROADCAST & WAIT "Stop"). Then depending on its angle of approach, the robot will turn right (preferred mode) or left to avoid the obstacle, even though at this point the user may still be pushing one of the Arrow Keys (which is ignored by the robot). This autonomous evasion maneuver is achieved using a nested REPEAT UNTIL Structure as shown in the middle part of Fig. 4.34. The Outer REPEAT UNTIL Structure ensures that this escape maneuver will continue until both IR Sensors see "no obstacle" (i.e. when their values are smaller than "10"). The two Inner Parallel REPEAT UNTIL Structures ensure that the proper "turn direction" is taken, although "turning right" is preferred if the robot happens to be going straight and squarely into the obstacle.

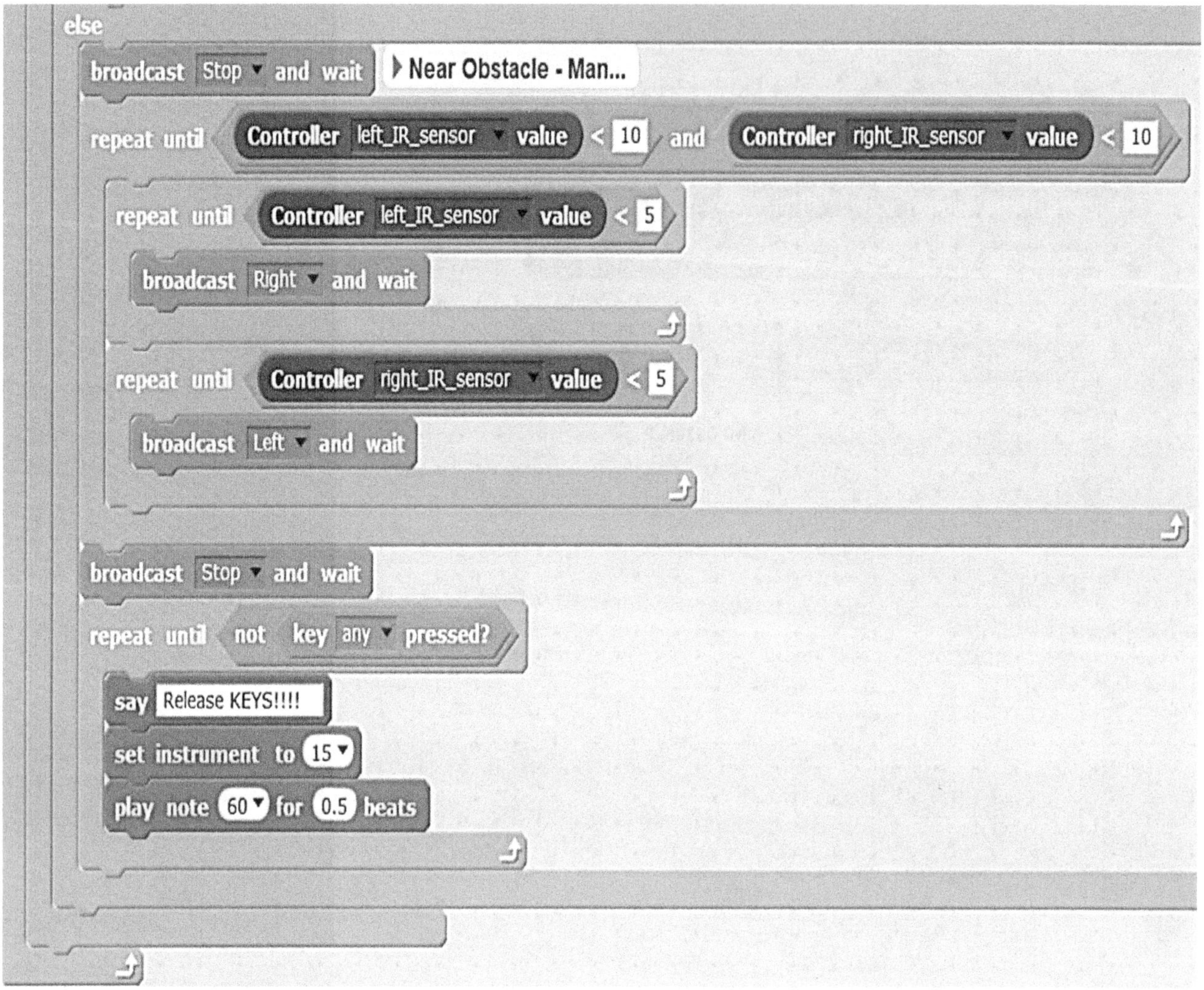

Fig. 4.34 Part 2 of MAIN FOREVER Loop in "SmartAvoider.sb2".

d) Eventually, the robot will be steered away from the obstacle, and it would then STOP. And at that point, if the user is still pushing on some key, the robot would keep on SAYing "Release KEYS!!!!" and playing Note 60 on Instrument 15 (Choir). When the user releases all keys (i.e. NOT ANY KEY PRESSED), the program is rerouted to the beginning of the MAIN FOREVER Loop (see Fig. 4.33), and the user regains the usage of the keyboard and the control of the robot.

Video 4.19 demonstrates the run-time performance of this program "SmartAvoider.sb2".

4.8 Line Follower – Obstacle Avoider

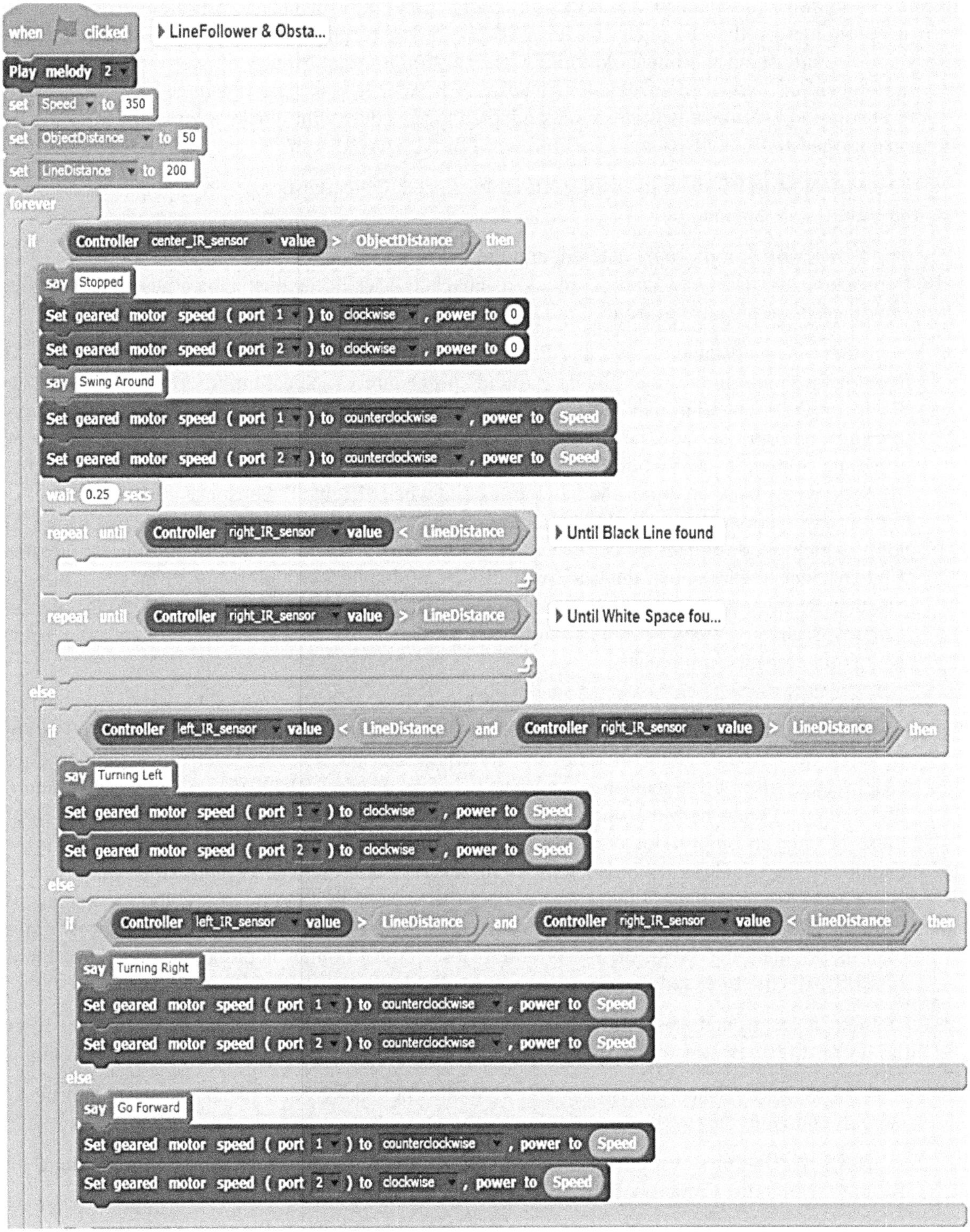

Fig. 4.35 MAIN SCRIPT in “LineFollower-ObstacleAvoider.sb2”.

In this section, we are going back to using the CarBot as described in Fig. 3.37 and we'll develop a SCRATCH solution equivalent to the TASK program "LineFollower-ObstacleAvoider.tskx" from Section 3.8. It is recommended that the reader reviews Section 3.8 to recall the behaviors of the Left and Right IR Sensors and to understand the threshold values used in the program "LineFollower-ObstacleAvoider.sb2". The same algorithm, described in detail in Section 3.8 (which won't be repeated here), is used for the SCRATCH solution, of course with the syntax adapted to the equivalent blocks allowed in SCRATCH. Fig. 4.35 describes the MAIN SCRIPT of "LineFollower-ObstacleAvoider.sb2":

1) The MAIN SCRIPT starts by setting Variables Speed, ObjectDistance and LineDistance respectively to the Values 350, 50 and 200.

2) The FOREVER Loop is next entered, and the PC reads in the latest "Controller Value" of the Center IR Sensor and use it in the First IF block to check this value against ObjectDistance, to figure out whether there is an obstacle in front of the CarBot or not:

- If there is an obstacle present, the PC stops the bot by turning off power to the 2 Geared Motors. Next it says and initiates the "Swing Around" procedure which essentially rotates the CarBot to the Right for 0.25 second to make sure that the Right IR Sensor is off the black track and on top of the background paper surface (i.e. IR-Right's Value is now greater than LineDistance). At this point in time and in the physical world, we would want the CarBot to keep on rotating right until the Right IR Sensor gets back on top of the black track (i.e when "Right-IR-Sensor" < "LineDistance") and in the other direction of travel. This logical condition is implemented by the First REPEAT UNTIL loop which is designed to be executed if the most recent IR-Right reading is greater than LineDistance, meaning that the CarBot is still seeing the background paper surface and therefore the "Rotate Right" motion should be maintained. This loop effectively does "nothing" because the CarBot is already doing what it is supposed to do, i.e. rotating right (please note the empty slot inside this Repeat Until block).

- Eventually, the CarBot crosses over the black line, and IR-Right's Value would be less than LineDistance, and the First REPEAT UNTIL loop is exited and the Second REPEAT UNTIL loop is taken up. At this point in time, the IR Right Sensor sees the black track, so the Condition (Right_IR_Sensor > LineDistance) is FALSE, so the Second REPEAT UNTIL loop is entered and maintained. Please note that the CarBot is "already doing the right thing", i.e. rotating right, thus there is only an empty slot (i.e. "do nothing") inside the Second REPEAT UNTIL loop. But eventually the CarBot's Right IR Sensor crosses over into the background paper area, then "Right-IR-Sensor" is larger than "LineDistance" and the Second REPEAT UNTIL loop is exited. At that time, the CarBot is physically straddling the black track but oriented in the opposite travel direction than its original one. Next, the PC program flow control pointer is passed to the end of the overall IF-ELSE-IF structure, and the algorithm restarts the FOREVER LOOP once more.

3) If there is no frontal obstacle, the ELSE branch of the First IF would be taken up, whereas a continuing IF-ELSE-IF structure is used to determine three mutually exclusive condition/action pairs:

- IF IR-Left sees Black and IR-Right sees White (i.e. background paper surface), then the CarBot should rotate into the Left to align the CarBot again with the Black Track.

- ELSE, IF IR-Left sees White and IR-Right sees Black, then the CarBot should rotate into the Right to realign the CarBot with the Black Track.

- Otherwise, i.e. the Final ELSE, the CarBot should just drive Forward. At that time, the CarBot could be right on top of the Black Track or it could be right on top of the Background Paper (i.e. it is off-track). This uncertainty is unavoidable due to the placements of the Right and Left IR Sensors with respect to the width of the Black line (see previous discussions in Section 3.8).

Please see Video 4.20 for the runtime performance to this program "LineFollower-ObstacleAvoider.sb2". It is not as responsive as the TASK version (Video 3.25) but much better than the PLAY700 (CM-50) version also in SCRATCH 2 (https://youtu.be/lwpsJu5uRu0).

4.9 Dowel Scanner

In this section, we are going back to using the Dowel Scanner robot described in Fig. 3.53 and we'll develop a SCRATCH solution equivalent to the TASK program "DowelScanner.tskx" from Section 3.9.2. It is recommended that the reader reviews Section 3.9.2 to recall the behaviors of the Servo Motor (Port 3) and the Frontal IR Sensor (Port 4) and to understand the various Parameter values used in the program "LineFollower-ObstacleAvoider.tskx". The same algorithm, described in detail in Section 3.9.2 (which won't be repeated here), is used for the SCRATCH solution, of course with the syntax adapted to the equivalent blocks allowed in SCRATCH. Fig. 4.36 describes Part 1 of the INIT Event in "DowelScanner.sb2":

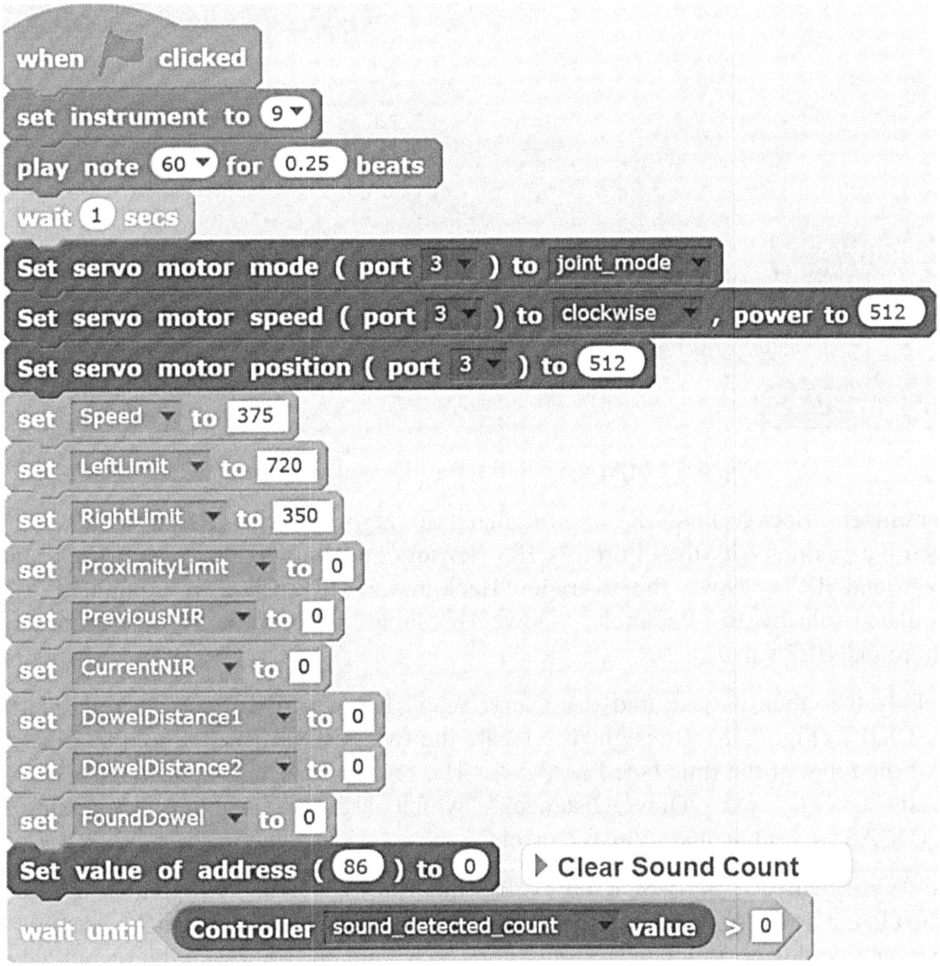

Fig. 4.36 Part 1 of INIT Event in "DowelScanner.sb2".

1) First Note 60 is played on Instrument 9 for 0.25 beat, and the PC is waiting on this action for 1 second.

2) The Servo Motor on Port 3 is set to Joint Mode, its Speed (Power) is set to 512 and it is sent to Position 512,

3) Next a series of Parameters are set to their initial values, from “Speed” to “FoundDowel”.

4) The special command “Set Value of Address 86 to 0” essentially initializes the “Sound Detected Count” to zero. The reader needs to consult the file named “CM-150_e_Manual.pdf” (included with the Example Codes ZIP file) and to search for this entry in the Control Table towards the end of this file. The next WAIT UNTIL block just waits for the user to clap his or her hands.

5) Once the robot hears the user’s clap, the MAIN SCRIPT is executed (see Fig. 4.37):

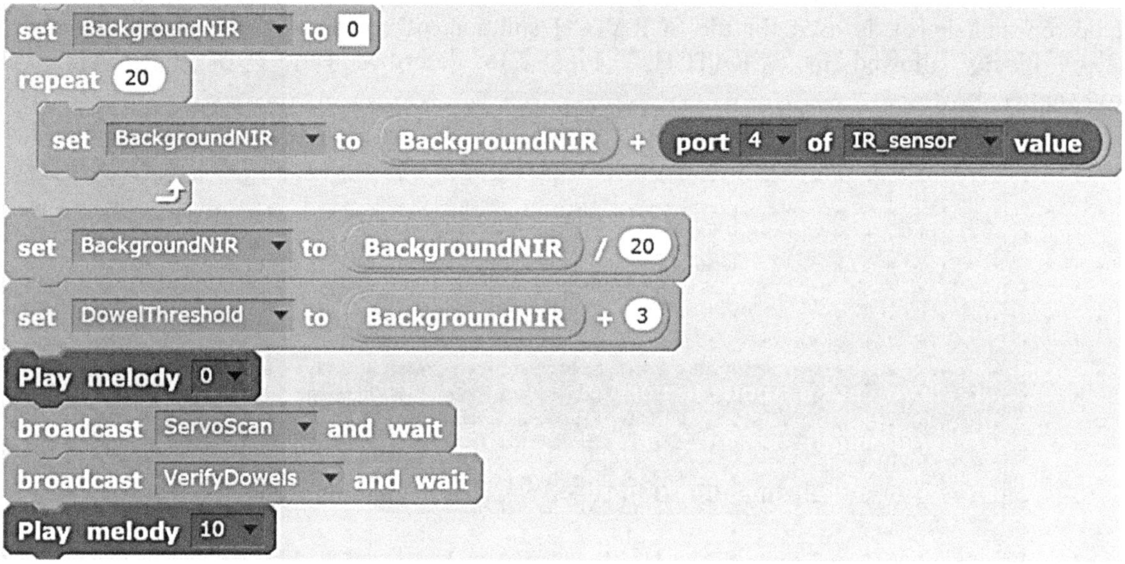

Fig. 4.37 MAIN SCRIPT for “DowelScanner.sb2”.

- Parameter “BackgroundNIR” is initialized to zero, and the REPEAT-20 Loop collects 20 consecutive values of the Port 4 IR Sensor and sums them up back into Parameter “BackgroundNIR”. Next, the average “BackgroundNIR” value is computed by dividing the accumulated sum by 20. Parameter “DowelThreshold” is set to be 3 units larger than the average “BackgroundNIR” value.

- Melody 0 is then played, and the “ServoScan” Event is BROADCASTed. The “ServoScan” SUB-SCRIPT (Fig. 4.38) is designed to locate the two dowels which would be placed at random in front of the robot at run time (see Fig. 3.54). The results from “ServoScan” are the two Parameters “DowelDistance1” and “DowelDistance2” which are used in the processing of the next BROADCAST Event named “VerifyDowels”.

- The “VerifyDowels” SUB-SCRIPT is illustrated in Fig. 4.40. It is written to allow the robot to find the closest dowel and approach it within a given distance, next it waits 5 seconds for the user to remove the first dowel, and then the robot goes on to find and approach the second dowel.

- Finally, the robot plays Melody 10 and the “DowelScanner” program terminates.

6) Details for the "ServoScan" SUB-SCRIPT are described in Fig. 4.38:

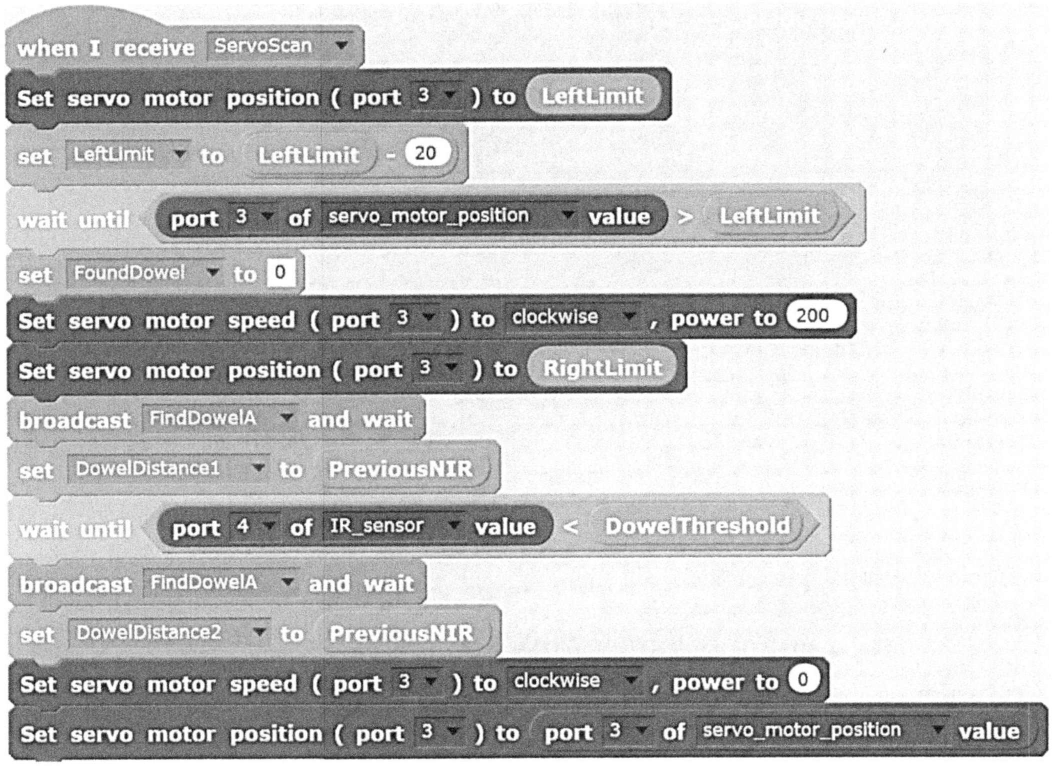

Fig. 4.38 SUB-SCRIPT "ServoScan" in "DowelScanner.sb2".

• First the Servo Motor on Port 3 is set to go its LeftLimit Position while the PC waits on the Servo Motor to get there.

• Parameter "FoundDowel" is cleared to zero and power (i.e. speed) to the Servo Motor is reduced to 200. Then the Servo is sent on its way to its RightLimit, and because of the lower speed used, the robot will take quite a bit of time to get there. It is during this time that the robot needs to figure out the "NIR" distances to the First Dowel (DowelDistance1) and the Second Dowel (DowelDistance2). For each dowel, the robot BROADCAST the Event "FindDowelA" to find the corresponding dowel distance.

• The details of Event "FindDowelA" are described in Fig. 4.39. At the beginning of the search for the first dowel, the Port 4 NIR Sensor "sees" empty space, thus the WAIT UNTIL block is written to allow the PC to "do nothing" until the NIR Sensor starts to detect the left edge of the first dowel (i.e. when IR Sensor > DowelThreshold). Then, "PreviousNIR" is set to the value found in "BackgroundNIR", and "CurrentNIR" is set to the latest IR Sensor value. The two parameters are "seeded" this way to ensure that logically "NIRDifference" will have a positive value. Therefore, the upcoming REPEAT UNTIL Loop will be entered for sure.

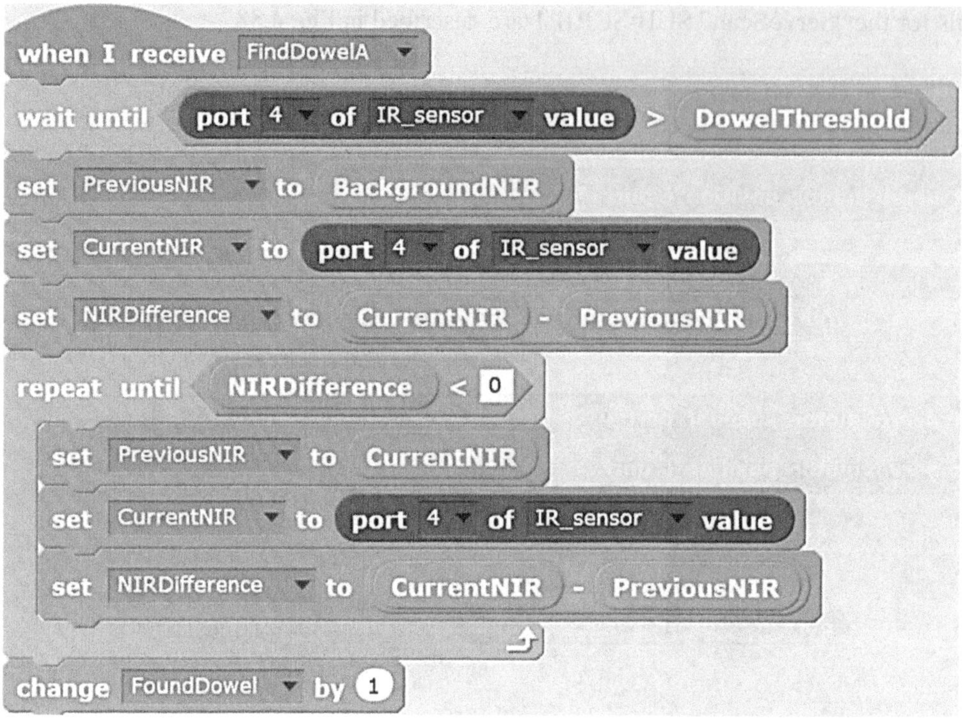

Fig. 4.39 SUB-SCRIPT "FindDowelA" in "DowelScanner.sb2".

• The REPEAT UNTIL loop is designed to find the approximate center line of the round dowel (if needed, please review Fig. 3.55). This search algorithm essentially swaps the "CurrentNIR" value into "PreviousNIR", and the next "CurrentNIR" is updated with the latest reading from Port 4. Between the dowel's left edge and its center line, the IR Sensor readings will increase and therefore "NIRDifference" (= CurrentNIR – PreviousNIR) will be positive during the same time. This ensures that the REPEAT UNTIL Loop is maintained. As soon as the dowel's centerline gets passed over, the IR Sensor readings will decrease, making (NIRDifference < 0), and the loop terminates.

• Finally, "FoundDowel" gets incremented by "1" and the Event "FindDowelA" ended for the first dowel. When Event "FindDowelA" gets called for the second time by Event "ServoScan" which is searching for the second dowel, the previous steps are executed and "FoundDowel" gets incremented to "2".

• Now we can get back to the last two blocks found in Fig. 4.38 which effectively stop the Servo Motor in its track. At this point, the Servo Motor should be aiming towards the second dowel (i.e. the "Right" dowel). Next, the SUB-SCRIPT "VerifyDowels" starts.

7) Details for the "VerifyDowels" SUB-SCRIPT are described in Fig. 4.40:

• Once the previous SUB-SCRIPT "ServoScan" is done (Fig. 4.38), Parameters "DowelDistance1" and "DowelDistance2" are assigned with their actual values. SUB-SCRIPT "VerifyDowels" now takes over and it triggers an appropriate sequence of Event-BROADCASTs ("CenterSensor", "Right", "Left", and "FindApproachDowel") depending on the actual runtime relationship between "DowelDistance1" and "DowelDistance2". For example, if "DowelDistance2" is larger than "DowelDistance1", this means that the Right Dowel is closer to the robot than the Left Dowel, thus the robot first straightens up the NIR Sensor forward (Event "CenterSensor") and then turns right towards the Right Dowel (Event "Right"). Once the robot finds the Right Dowel with the NIR Sensor, it approaches this dowel and when it is close enough, it will play Melody 3 signaling to the user to remove the Right Dowel (Event "FindApproachDowel"). Next it swings to the Left looking for the Left Dowel (Event "Left") and performs similar tasks of finding and approaching the Left Dowel (Event "FindApproachDowel).

• If the other situation happens instead (i.e. the ELSE branch), the Left Dowel is closer to the robot than the Right Dowel, then the robot would execute similar find-approach tasks for the Left Dowel first, then turns right to take care of the Right Dowel.

Fig. 4.40 SUB-SCRIPT "VerifyDowels" in "DowelScanner.sb2".

8) Details of SUB_SCRIPT "CenterSensor" are shown in Fig. 4.41, whereas Port 3 Servo Motor is sent to Position 512 at Speed 512, and the PC waits 0.5 second for the Servo to get there.

Fig. 4.41 SUB-SCRIPT "CenterSensor" in "DowelScanner.sb2".

9) Details of SUB-SCRIPT "FindApproachDowel" are shown in Fig. 4.42:

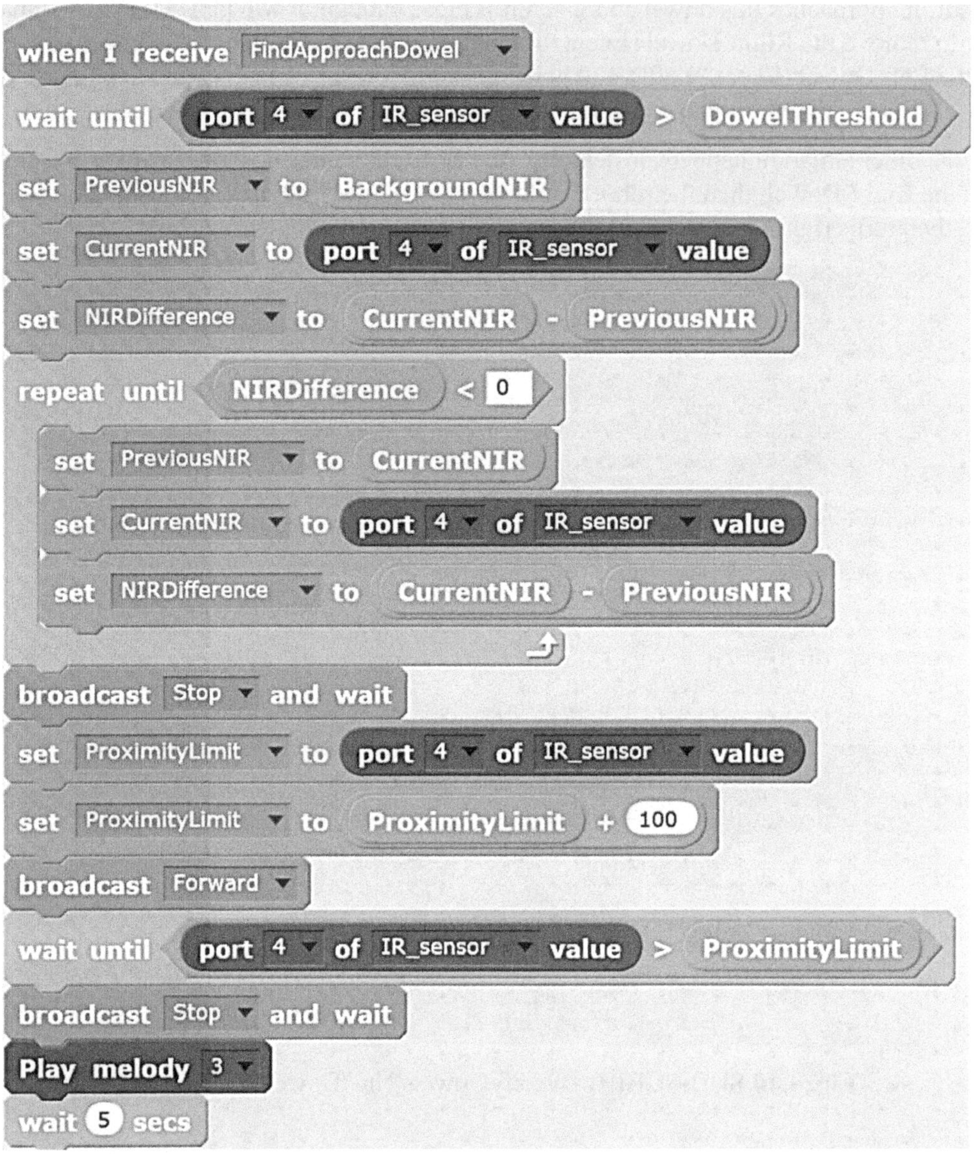

Fig. 4.42 SUB-SCRIPT "FindApproachDowel" in "DowelScanner.sb2".

- The reader should note that the robot was already rotating Left or Right when Event "FindApproachDowel" is triggered. Thus, the first WAIT UNTIL Loop is designed for the robot to "do nothing" until the Port 4 IR Sensor starts to "see" the edge of the dowel being sought for. When this occurs, "Port 4 IR Sensor" > DowelThreshold, then the SCRATCH code jumps to the same dowel search algorithm previously explained and implemented by the next three SET blocks (PreviousNIR, CurrentNIR and NIRDifference) and the REPEAT UNTIL Structure.

- Once the dowel is found, the robot STOPs, reads in the latest IR Sensor Value and saves it into Parameter "ProximityLimit".

- Parameter "ProximilyLimit" is next increased by 100 units, and the robot is sent Forward.

• Next, a WAIT UNTIL block monitors the Port 4 IR Sensor reading to see if it exceeds the current "ProximityLimit" (please note that the robot is still moving forward towards the "found" dowel). When the robot gets close enough to the dowel to make (IR Sensor Value > ProximityLimit), then it is time to STOP the robot and play Melody 3. Lastly, the PC/Robot waits for 5 seconds for the user to remove the "verified" dowel, before the SCRATCH code gets back to its appropriate place in the SUB-SCRIPT "VerifyDowels" (Fig. 4.40).

Video 4.21 shows the runtime performance of this "DowelScan" SCRATCH project using USB "wired" and BlueTooth wireless communications. This video shows that robotics tasks that require high speed synchronization between sensors and actuators cannot be implemented satisfactorily using the R+SCRATCH-SCRATCH2 environment due to the unavoidable communications delay between the PC and the CM-150 firmware controlling the sensors and actuators on the robot.

Recapping Chapters 1 through 4, we can draw the following conclusions:

1) The "Sense-Think-Act" paradigm and the Condition/Action analysis procedure had been helpful in the beginning steps of understanding a given robotic challenge and to sort out the relevant Condition/Action pairs.

2) The actual software tools (TASK or SCRATCH) we use to solve this challenge may produce different runtime performances on the same physical robot and may even lead us to have to change the original goals of the challenge.

3) If the robotic challenge requires rather quick actuator responses or fast changing environmental inputs, then TASK is a more suitable tool to use.

4) The SCRATCH's EVENT PROGRAMMING techniques of BROADCAST and BROADCAST & WAIT are quite powerful, but unfortunately the robotics-oriented example codes in this Chapter 4 did not illustrate their full potentials as they were only used to "mimic" TASK functions using BROADCAST & WAIT. Please see Exercise 12 also for a very important side effect to resolve when using BROADCAST techniques with the CM-150 system.

4.10 References

Bagge, P (2015) How to teach Primary Programming using SCRATCH. University of Buckingham Press, Buckingham UK.

Ford, JL Junior (2014) Scratch 2.0 programming for teens. CENGAGE Learning PTR, Boston USA.

Vlieg, EA (2016) Scratch by Example: Programming for All Ages. Apress Media, LLC, New York USA

Warner, TL (2015) Teach yourself Scratch 2.0 in 24 hours. SAMS, Indianapolis USA.

4.11 Review Questions for Chapter 4

1. Currently, which ROBOTIS controllers does the software tool R+SCRATCH support?

2. How many CM-50 Block Codes are available in the ROBOTIS Extension Package for the SCRATCH 2 IDE?

3. Describe the two types of the CM-50 Extension Block Codes available in the SCRATCH 2 IDE.

4. How long does it take for R+SCRATCH to process a "r" CM-50 Block Code?
5. How long does it take for R+SCRATCH to process a "w" CM-50 Block Code?
6. Describe the differences between "BROADCAST" and "BROADCAST & WAIT".
7. Describe the differences in processing keyboard strokes (arrow keys) between SCRATCH 2 and TASK tools.
8. Give an example when the ROBOTIS's "Load Current Value Of" block code has to be used.
9. Which SCRATCH 2 Block Code needs to be used when ones want to display a text message by the Sprite in the Stage Area?
10. Describe how to use a "Musical Instrument" inside the SCRATCH 2 IDE.
11. What is the critical step when using the REPEAT UNTIL loop structure inside the SCRATCH 2 IDE?
12. The program "IR-Based-Maneuvers-BW.sb2" uses the "Broadcast and Wait" approach. Try this program with and without the block "SAY IR_Sensor"? Discuss the results obtained.

Chapter 5: Mechanical Design Concepts in Level 1 Robots

The parts in the DREAM II kits are designed to mimic standard mechanical parts found in everyday products, e.g. plates, links, rivets, rods, etc… (see Fig. 5.1 for a "Squirrel on Skis"). Thus, there are many sources of information on mechanical engineering and design that the readers can draw from.

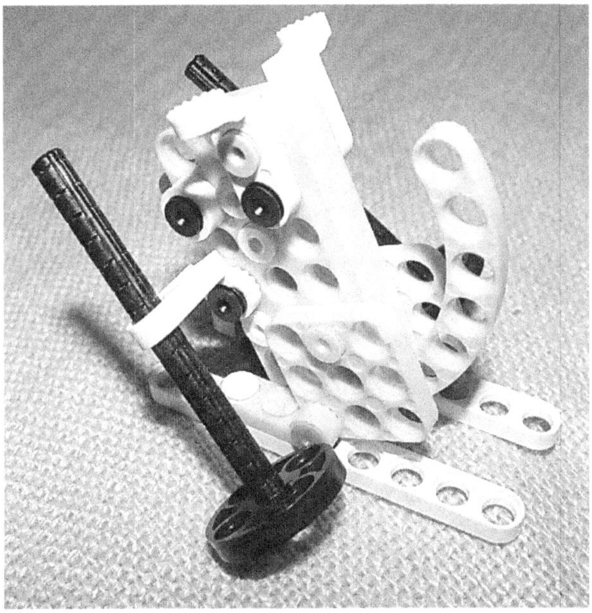

Fig. 5.1 "Squirrel on Skis".

ROBOTIS has its own DREAM II materials but some of their web materials are currently available only in Korean (Spring 2018), thus the reader will need to use Chrome as web browser so that it can translate them into "useable" English:

1) ROBOTIS DREAM II materials are at (http://support.robotis.com/ko/product/education/dream2.htm).

2) At present, the R+DESIGN tool has 3D CAD mechanical designs for up to the DREAM I kit only, but the materials for the DREAM II kit will be incorporated into this tool in near future. R+DESIGN can be downloaded at (http://www.robotis.us/roboplus2/) and its user manual can be viewed at (http://emanual.robotis.com/docs/en/software/rplus2/design/). R+DESIGN can be used to design and assemble your own robot design using existing DREAM kits parts.

Cornell University has a comprehensive library of mechanisms at its KMODDL web site (http://kmoddl.library.cornell.edu/). For example, if the reader wants to know more about the 4-bar linkage used for the Ladybug robot, this web page is quite comprehensive (http://kmoddl.library.cornell.edu/).

There are other numerous web sites for mechanism design and simulation, the author is listing just a few below:

http://www.mekanizmalar.com/

https://www.youtube.com/user/thang010146/videos

https://www.youtube.com/playlist?list=PLhoXNQqrCmEfAaTf0AfQ1Ztxmz2DoZiCk

http://www.roboanalyzer.com/mechanalyzer.html

There should be numerous books about "mechanisms" and "mechanics of machines" if the reader requires more formal references on the web or at your local libraries.

5.1 Direct Motor Drive

In Level 1, the Helicopter is an example of using the Geared Motor directly to move a device such as the 4-bladed rotating wing (see Fig. 5.2).

Fig. 5.2 Direct Drive of Helicopter's Rotating Wing.

5.2 Gear Applications

Technically speaking, the DREAM kit only provides sprocket type of parts (2 sizes – SPS-4 and SPS-17) that can be used to obtain spur-gear functionality (https://en.wikipedia.org/wiki/Gear). Fig. 5.3 shows a typical gear train that can be constructed with these sprocket parts. Although there is quite a bit of "free play" between the teeth of contacting sprockets, overall there is conservation of kinematic gear ratio: for example, in Fig. 5.3, one turn of the right large sprocket yields one turn of the left large sprocket and in the same direction.

Fig. 5.3 Parallel Gear Train constructed from DREAM kit Sprocket Parts.

Fig. 5.3 is an application where all 3 sprockets' axes of rotation are parallel to each other, but they also can be used in designs where these axes are rotated 90 degrees from each other in a crossed configuration (see Fig. 5.4).

Fig. 5.4 Crossed Gear Train constructed from DREAM kit Sprocket Parts.

The sprockets can also be stacked to provide a more compact gear reduction system as shown in Fig. 5.5.

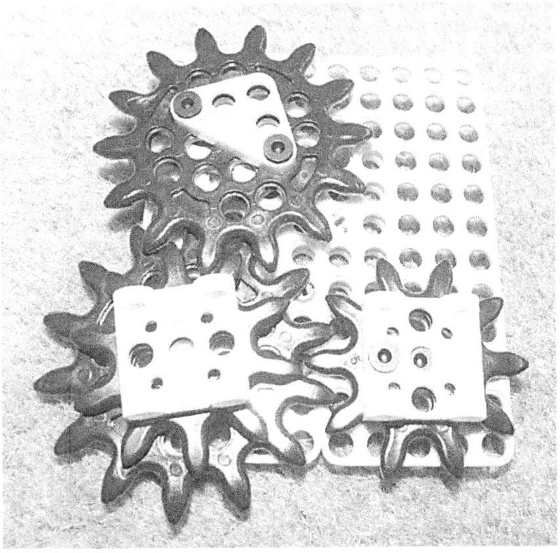

Fig. 5.5 Stacking Sprockets to achieve a compact gear reduction system.

5.2.1 Using Sprockets (Windmill & Dragonfly)

The Windmill robot combines both parallel and crossed axes designs using 4 sprockets (see Fig. 5.6), while the Dragonfly robot (see Fig. 5.7) uses the same gear reduction system illustrated in Fig 5.5.

Fig. 5.6 Gear Train used in Windmill Robot.

Fig. 5.7 Gear Train used in Dragonfly Robot.

Later in Level 3, a Gear Train similar in design to the one shown in Fig. 5.3 is used to obtain a 4-wheel drive CarBot (see Fig. 5.8). More discussions about this Robot are presented in Section 6.9.

Fig. 5.8 Four-Wheel Drive CarBot from DREAM II Level 3.

5.2.2 Using Pulleys & O-rings (i.e. Tires)

Ones can also obtain "gear functionality" using pulleys and O-rings, i.e. tire systems (see Fig. 5.9), but this system is not as mechanically efficient as the sprocket-based one, as there can be more slippage between the tires contact surfaces. Please note that the big tire on the right comes only from the Bioloid STEM kit.

Fig. 5.9 Gear Functionality using Tire Systems.

5.3 Linkage Applications

To extend the usefulness of Geared and Servo Motors beyond their basic rotational motions, linkage systems can be used (https://en.wikipedia.org/wiki/Linkage_(mechanical)). In the DREAM II system, 4-bar linkage systems and their variants are used to obtain various walking motions. More sophisticated linkage systems have been developed for robot walking motions such as the Klann and Jansen systems (https://en.wikipedia.org/wiki/Klann_linkage; https://en.wikipedia.org/wiki/Jansen%27s_linkage).

5.3.1 Whale

Real whales (and dolphins) swim forward by "flexing" their bodies up and down. This motion is mimicked (as much as possible) with the mechanism shown in Fig. 5.10, whereas the rotational motion of a Geared Motor is transmitted via the Horn/Crank to the Connecting Rod whose tip is constrained to trace an up-down and forward-backward trajectory because of the restricting box frame.

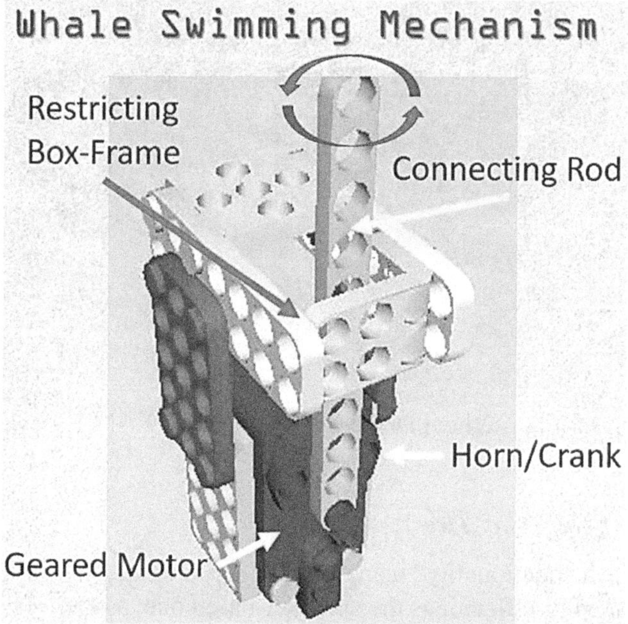

Fig. 5.10 Whale's Swimming Mechanism.

This "swimming" mechanism derives from a slider-crank mechanism (see Fig. 5.11), but without the slider part. The slider-crank mechanism itself is a special case of the 4-bar linkage system (see Fig. 5.12) which is often applied in robot design for the DREAM II series (https://en.wikipedia.org/wiki/Four-bar_linkage).

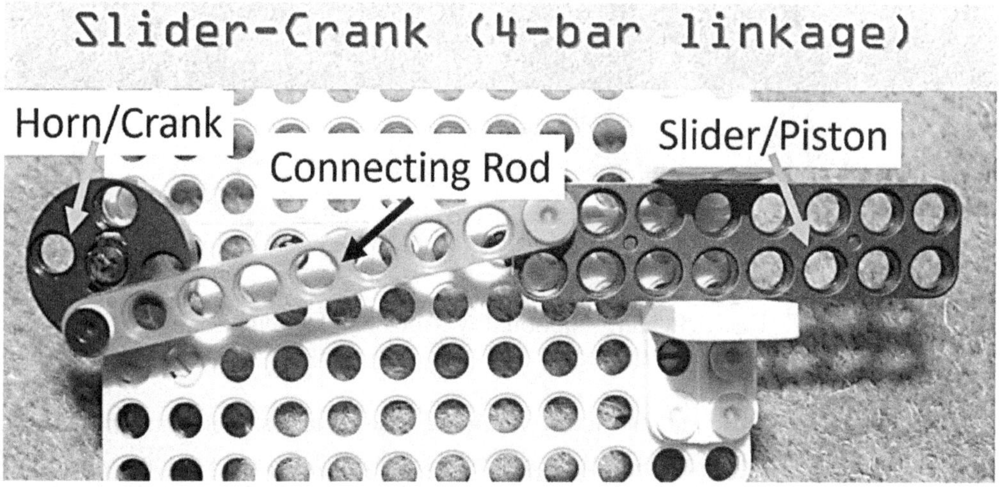

Fig. 5.11 Slider-Crank Mechanism.

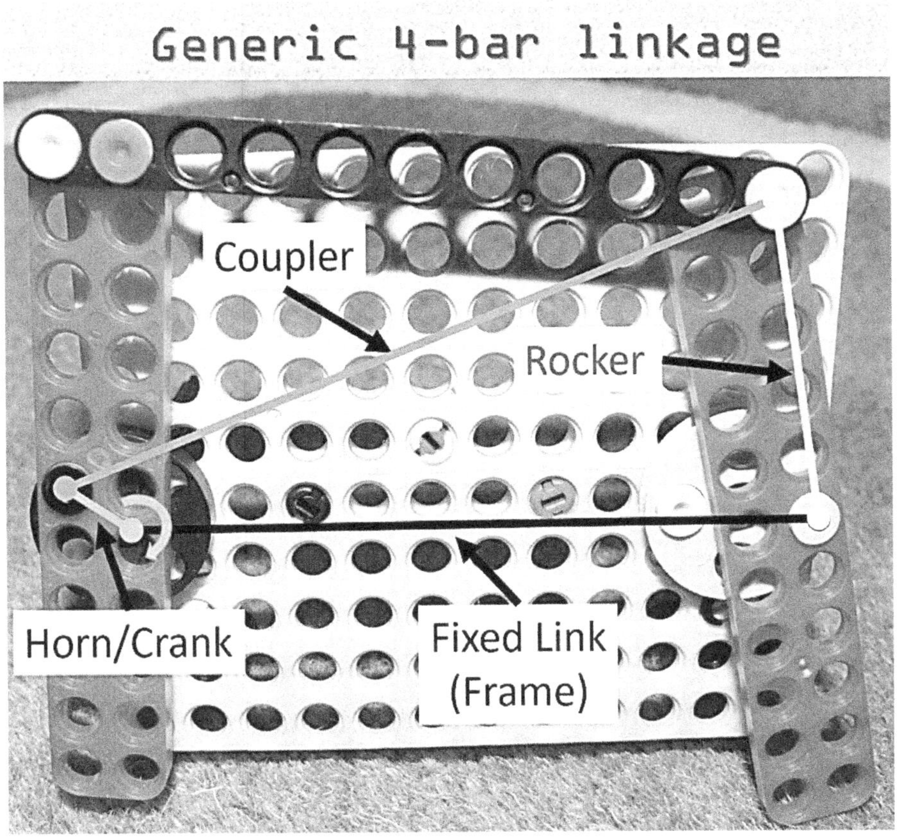

Fig. 5.12 Generic Four-Bar Linkage System.

Please visit this web site (http://www.mekanizmalar.com/fourbar.html) for an animation view of the working of a 4-bar linkage system whereas the reader can vary the lengths of the various linkages and see the effects on the whole linkage system.

Fig. 5.12 illustrates the 4 types of linkages used and their functions within a typical 4-bar system using parts from the DREAM kit:

1) The FIXED link does not move (relative to the robot) as it is always mounted onto the robot's body (i.e. main frame).

2) The CRANK link is usually the motor's horn which can fully rotate 360 degrees as it is attached to the motor's armature.

3) The ROCKER link of course has a back-and-forth motion which translate into a "walking" motion for the whole robot. Usually the ROCKER link corresponds to 1 physical part of the DREAM kit.

4) The COUPLER link is usually the hardest one to identify, as it is often created by linking two or more DREAM parts together. In Fig. 5.12, the COUPLER link is composed of the Left Red link (2x11 plate), rigidly connected via 2 rivets (top left corner) to the Top Black link (1x11 plate). Thus, very often, the COUPLER link when drawn does not coincide with a single physical part (as shown in Fig. 5.12). The COUPLER also has a back-and-forth motion like the ROCKER's.

The web page (https://en.wikipedia.org/wiki/Four-bar_linkage) also lists the required mathematical relationship between the lengths of the 4 links to have a functional 4-bar linkage system.

5.3.2 Brachiosaurus & Cow

Although physically, the Brachiosaurus and Cow robots look different from each other, kinematics wise both are applications of the 4-bar linkage design (see Fig. 5.13). In this Figure, the author has already identified the various links with different color codes and the reader is encouraged to finish this link identification job by labeling them with their proper names. Please note that both robots walk by shifting their weights in a left-right manner, i.e. their left and right 4-bar systems are in opposite phases. Both robots can go forward or backward equally well. Please note that because the Brachiosaurus is taller in size than the Cow, the battery, its heaviest component, is hung low to help with stability for the Brachiosaurus.

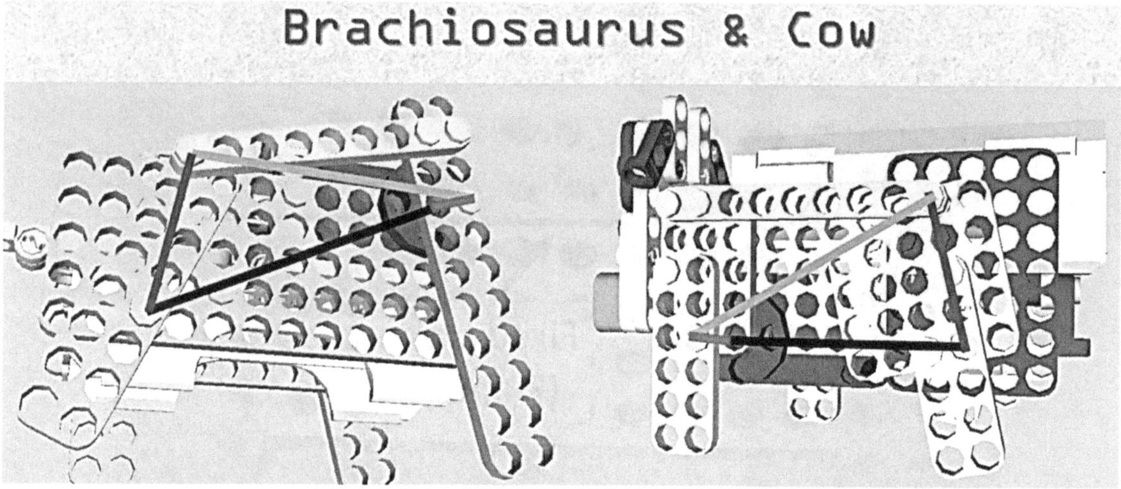

Fig. 5.13 Walking Mechanisms for Brachiosaurus and Cow robots.

5.3.3 Rabbit

The Rabbit robot also applies the 4-bar linkage design for its 4 legs, but its left and right 4-bar linkage systems are synchronized (i.e. they are in phase). Additionally, the Rabbit has an inclined body position and its Center of Gravity is also positioned just in front of its rear legs, both features contribute to its ability to "hop" forward very well.

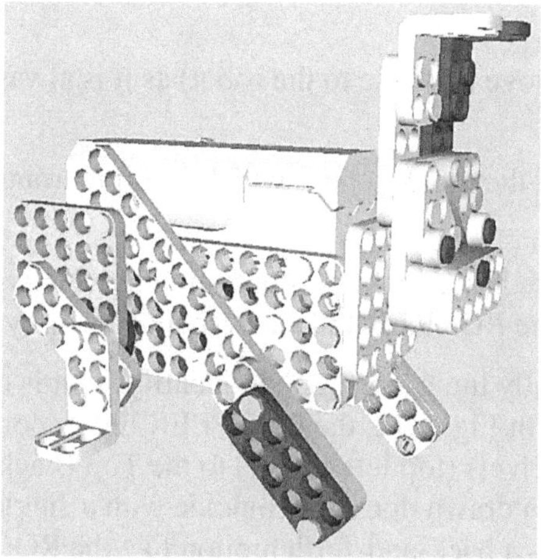

Fig. 5.14 The Rabbit's CG is positioned such that it can "hop" forward.

5.3.4 Ladybug

The LadyBug has 3 legs on each side of its body, and as a 4-bar linkage system can only provide for 2 walking legs, robot designers would have to merge two linkage systems together. Fig. 5.15 shows that the LadyBug has a 4-bar linkage system for its rear and middle legs, and a 5-bar linkage system for its front leg and middle leg (which is shared with the rear 4-bar linkage system). The crank link (red) is also shared between the rear and front linkage systems. Same as for the Brachiosaurus and Cow, the Right and Left walking mechanisms for the LadyBug are in opposite phase of each other.

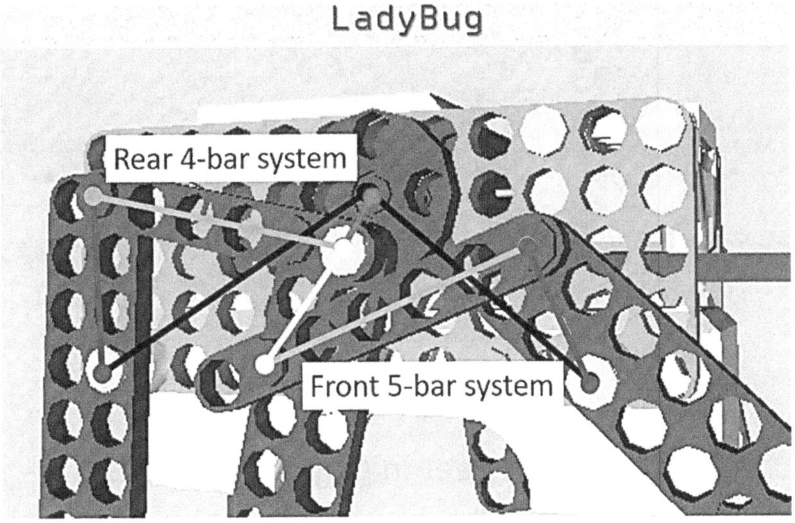

Fig. 5.15 The LadyBug's Rear and Front Linkage Systems.

For further illustrations, Fig. 5.16 shows three merged 4-bar linkage systems designed for a Scorpion robot (4 legs on each side) from the PLAY700 kit. For its two front legs, this Scorpio uses a Parallelogram Linkage whereas the Crank and Rocker links have equal lengths, likewise for the Coupler and Fixed links.

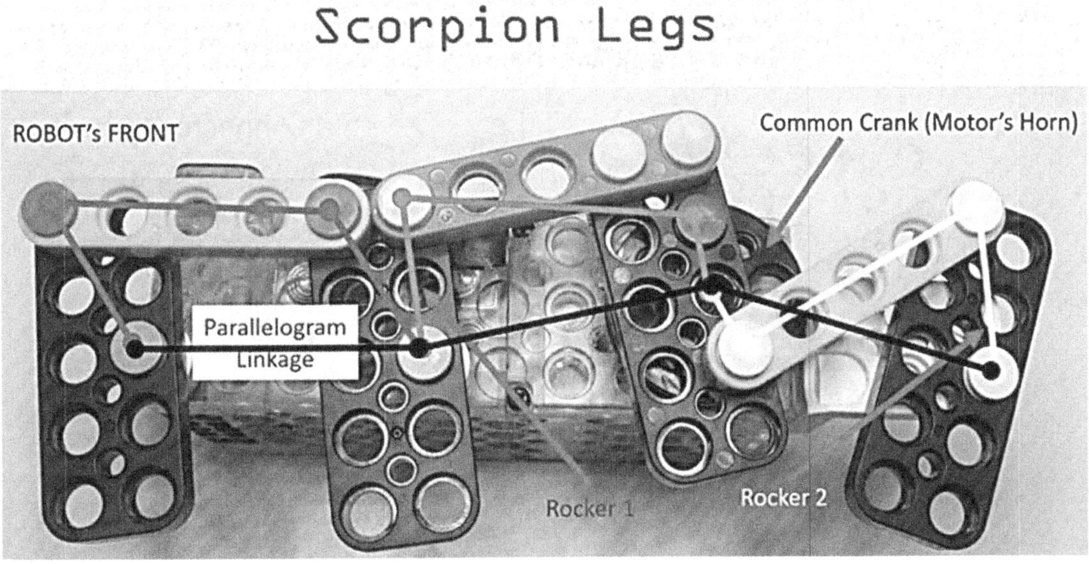

Fig. 5.16 The PLAY700 Scorpion linkage systems.

5.3.5 Chick

For the robots described so far, the walking mechanisms are independent for each side of the robot. The Chick robot design showcases a new approach where the left and right walking mechanisms are linked to each other via an elaborate median mechanical structure (e.g. its head – see Fig. 5.17).

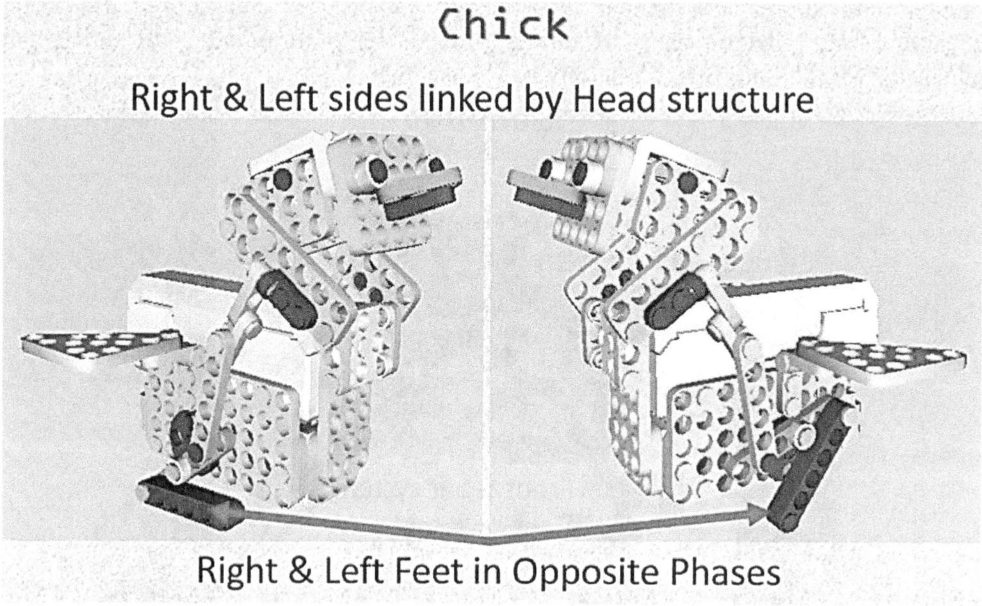

Fig. 5.17 Right and Left Views of the Chick robot.

The Chick also has its Right and Left walking mechanisms in opposite phase of each other (see Fig. 5.18 and please look for the positions of the crank on each view). The Chick's walking mechanism uses a Slider-Crank design, using the ground surface as the kinematic constraint.

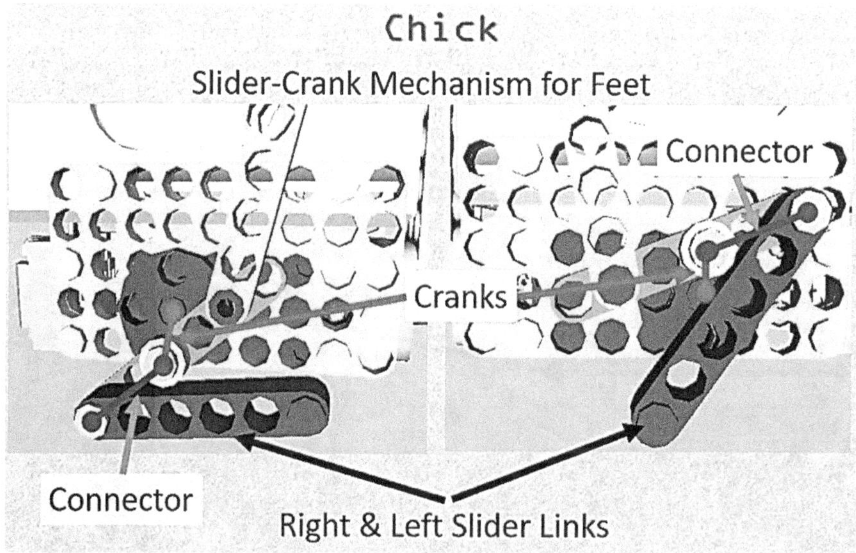

Fig. 5.18 Slider-Crank mechanisms used for the feet of the Chick robot.

In a way, we could consider the whole Head structure as being another 4-bar linkage with a complex Rocker link (see Fig. 5.19). This Rocker allows the Head to rock Left-Right as the Chick waddles on.

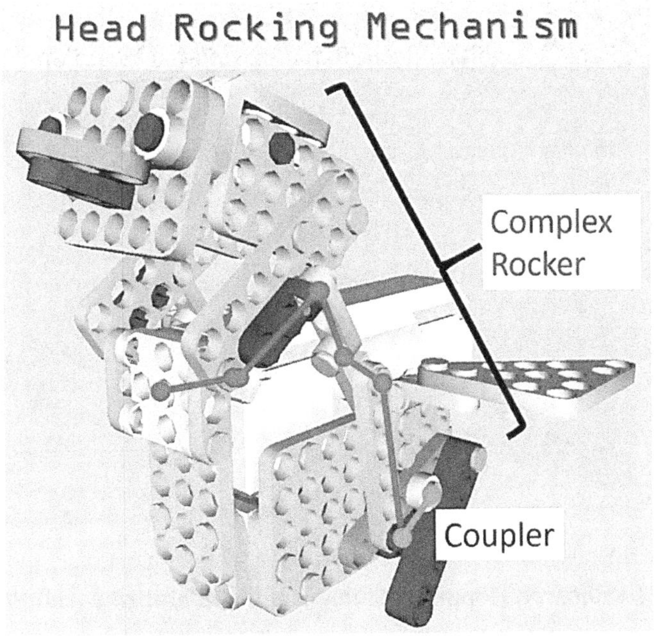

Fig. 5.19 Head Structure for the Chick robot.

Furthermore, Fig. 5.20 shows that there are complex "90 degree" reorientations of the rotation axes of the various links composing the Head structure. Please note that conceptually, the original 4-bar linkage system is a planar design and the Chick's Head structure illustrates 3-dimensional extensions to that idea.

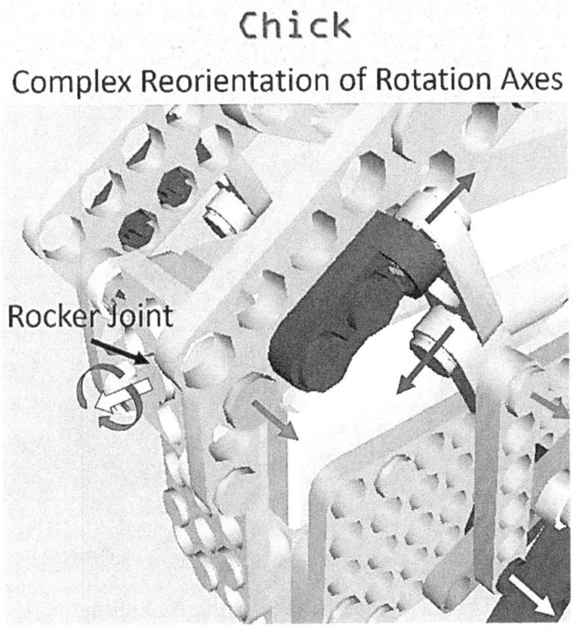

Fig. 5.20 Reorientations of rotational axes of link joints in the Chick's Head structure.

5.3.6 Kangaroo

Compared to the Chick robot, the Kangaroo robot has a simpler walking mechanism: it also uses a Slider-Crank design for its feet, but the Left and Right sides are synchronized (as the real Kangaroo can only hop around) and it uses its big tail to stabilize itself in the forward-backward direction (see Fig. 5.21).

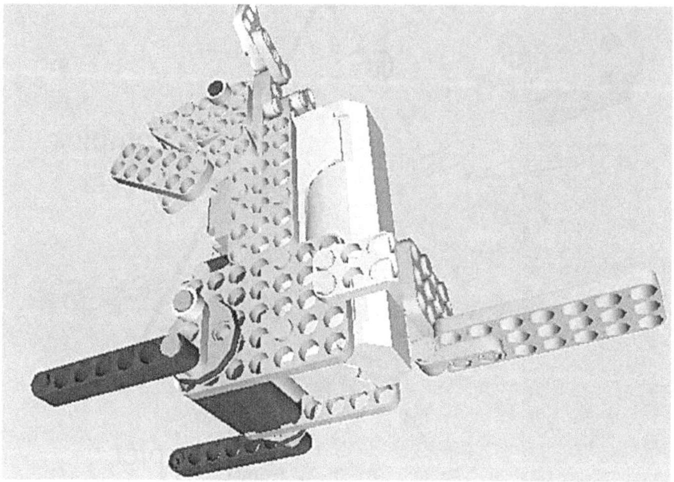

Fig. 5.21 Kangaroo Hopping Mechanism and Stabilizing Tail Structure.

5.3.7 Caterpillar

The Caterpillar robot mimics the motion of an Inch-Worm which alternates between body retraction and extension moves to achieve a forward (or backward) motion. The Caterpillar is designed with two distinct sections - front and rear, connected by two sets of linkages - top and bottom (see Fig. 5.22).

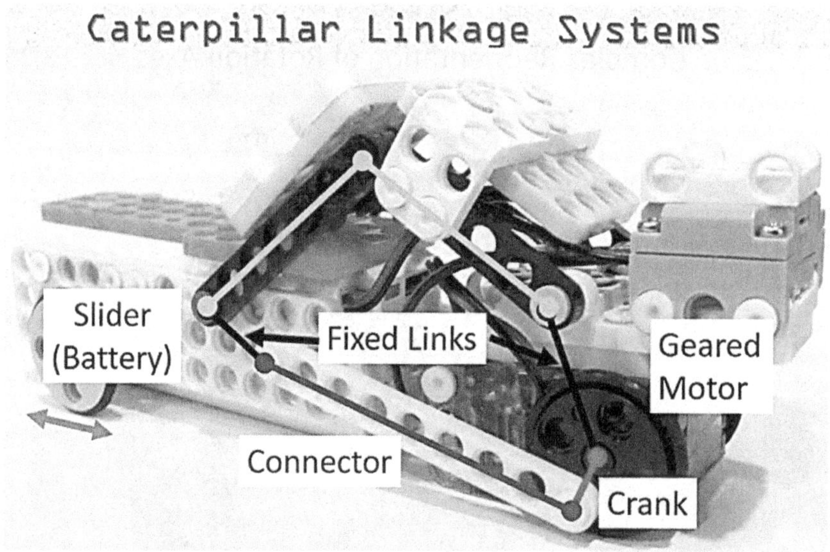

Fig. 5.22 Caterpillar Linkage Systems.

The Front Section has the Geared Motor and the Power Switch module, while the Rear Section has the LiPo Battery and Battery Case. Each Front or Rear Section has its own Fixed links (black links in Fig.

5.22). The Top Linkage system (2 green links) is purely passive and it is used to keep the Front Section down on the ground surface. The Bottom Linkage system is powered via the Geared Motor and it is of the Slider-Crank type. The Crank is the Motor's Horn as usual, and it uses the Connector link to transfer motion to the Slider in the rear. The Slider role is fulfilled by the Battery and Case combination which is also equipped with a pair of passive wheels to minimize contact friction with the ground surface which itself acts as a kinematic constraint.

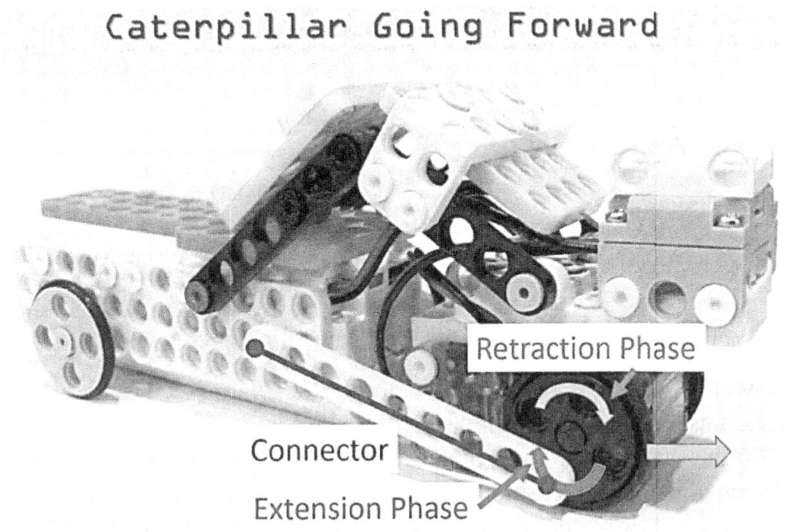

Fig. 5.23 Retraction and Extension phases for Caterpillar robot during Forward Motion.

When the Geared Motor (i.e. Front Section) is set to go Forward (see Fig. 5.23), during the Retraction Phase (i.e. Connector traveling on top half of the horn), the Rear Section is pulled forward, i.e. in the same direction as the Geared Motor. During this Retraction Phase, both Sections are working "together" and the user can see that the robot is moving forward. During the Extension Phase (i.e. Connector traveling on bottom half of the horn), the Geared Motor is still going forward, but it pushes the Rear Section away from itself (i.e. backward). During this Extension Phase, the Sections are working "against each other" and canceling out each other's motion, thus the robot would seem to be stopping overall during this time. A closer look would show that the Rear Section stops while the Front Section is pushed forward a small distance during this Extension Phase. Then the motion cycles back to the Retraction Phase.

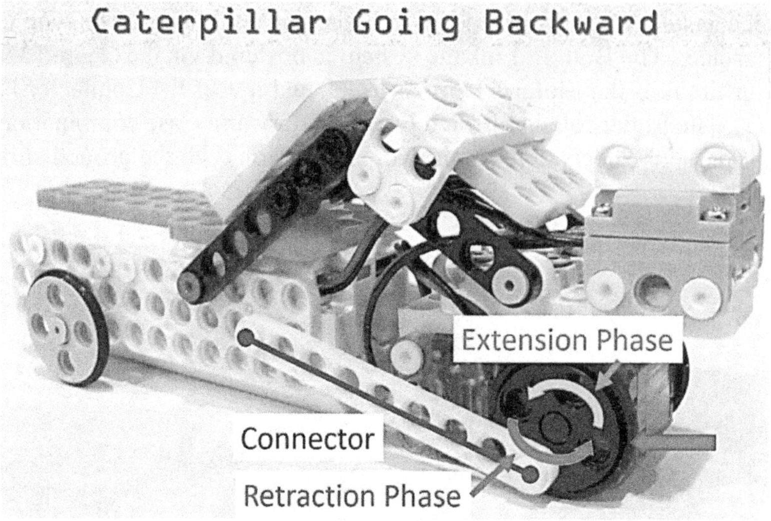

Fig. 5.24 Retraction and Extension phases for Caterpillar robot during Backward Motion.

When the Geared Motor (i.e. Front Section) is set to go Backward (see Fig. 5.24), everything gets reversed. During the Extension Phase (i.e. Connector traveling on top half of the horn), the Rear Section is pushed backward, i.e. in the same direction as the Geared Motor. During this Extension Phase, both Sections are working "together" and the user can see that the robot is moving backward. During the Retraction Phase (i.e. Connector traveling on bottom half of the horn), the Geared Motor is still going backward, so it wants to pull the Rear Section towards itself (i.e. forward). During this Retraction Phase, the Sections are working "against each other" and canceling out each other's motion, thus the robot would seem to be stopping overall during this time. A closer look would show that the Rear Section stops while the Front Section rolls backward a small distance during this Retraction Phase. Then the robot cycles back to the Extension Phase.

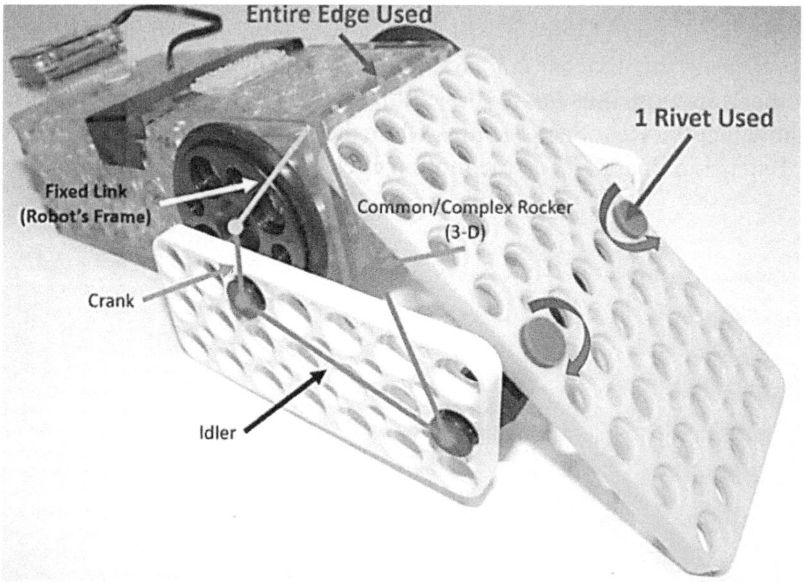

Fig. 5.25 Caterpillar robot from PLAY700 kit.

Fig. 5.25 shows another Caterpillar robot design using the PLAY700 kit. This one can go forward and backward like the DREAM version, but it can also turn left and right (please watch the YouTube video at https://youtu.be/tEK3LcaLrrl to see how it works).

Normally, the Rocker component is connected to the robot's frame via a fixed joint/node as in previous robot designs shown in this Chapter. However, with the PLAY700 Caterpillar, a structurally simpler, but more complex kinematics-wise, Rocker (i.e. yellow 5x9 plate) can "rock" and "slide" on the entire "top back edge" of the PLAY700 controller (see top of Fig. 5.25), i.e. to offer more "degrees of freedom" than the Chick's Head in Section 5.3.5.

5.3.8 Tyrannosaurus

The Tyrannosaurus also uses a 4-bar linkage system, but this linkage system is now used in the vertical direction (See Fig. 5.26). The Fixed Link is the robot main body as before, and the Crank Link is still represented by the Motor's Horn. The Rocker Link is now the Tyrannosaurus' arm (yellow coded parts). The Coupler Link is quite elaborate as it encompasses the big black foot, humongous thigh and connecting rod (all in orange coded parts in Fig. 5.26).

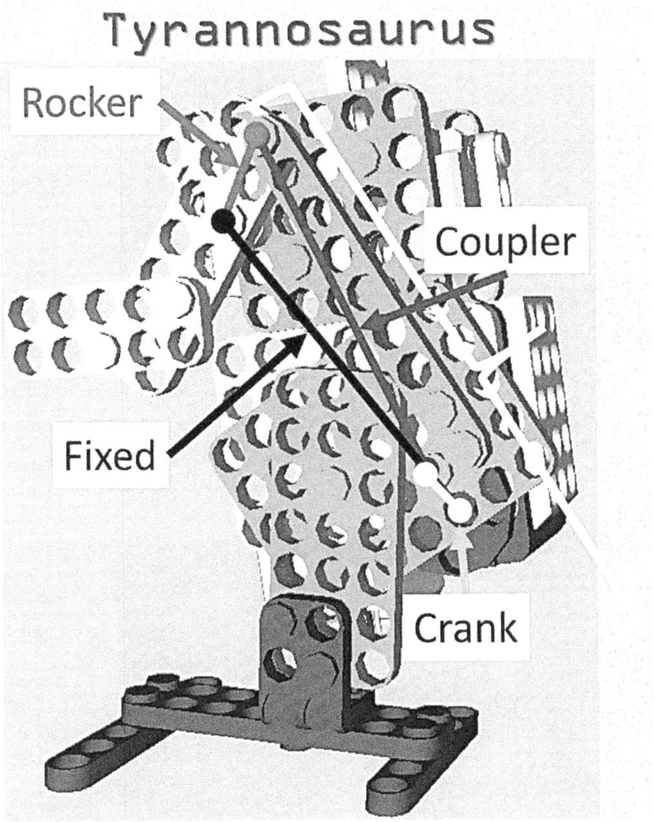

Fig. 5.26 Left 4-bar Linkage System used on Tyrannosaurus robot.

By design, the Tyrannosaurus robot uses its 4-bar Linkage Systems in opposite phases to achieve a Left-Right walking gait.

5.4 Review Questions for Chapter 5

Build your favorite linkage-based robots and:

1. Get out-of-phase walking bots to walk in-phase and vice-versa. Document their changes in behaviors.
2. Change rivet connection points on various linkage systems to see their effects, if any, on the overall workability of the modified systems.

Chapter 6: A Closer Look at Select DREAM II robots

The goal of this Chapter is to get the reader beyond the ROBOTIS-provided R+DESIGN example constructions and example TASK codes, by providing additional hardware designs as well as code enhancements whenever possible and appropriate. Most of these enhanced robots will require the user to obtain additional actuators and sensors beyond the ones included in the DREAM II School Set.

6.1 Elephant with NIR Sensor

In all the DREAM robots described so far, the reader should have noticed that the two Geared Motors (GM) were always mounted on the same axis of rotation and robot directional changes were made by modifying the rotation direction of one GM versus the other, and/or by varying the rotational speed of one GM with respect to the other. The Elephant design offers a rather unusual locomotion system whereas the two Geared Motors axes of rotation are set up perpendicular to each other instead (see Fig 6.1). One GM (Port 2) drives two small pulley-wheels (Part No. SPO-4X and SPO-5) to move forward and backward, while the other GM (Port 1) provides left and right directional changes using a Medium Tire (Part No. STR-39).

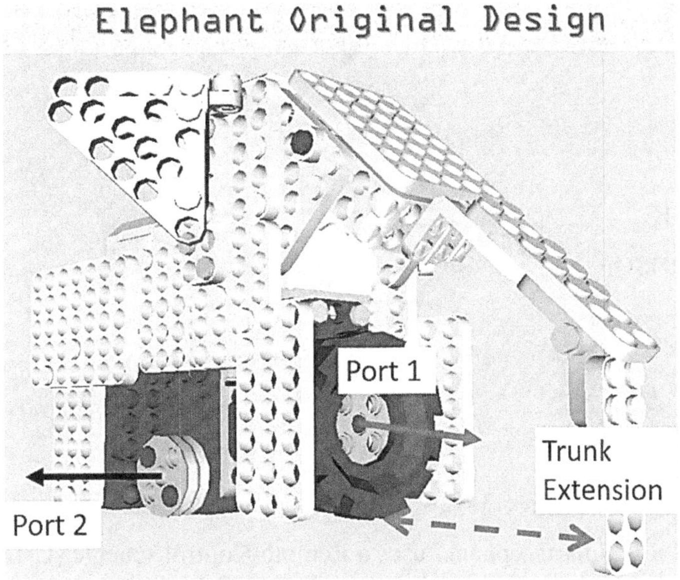

Fig. 6.1 Original Elephant Robot Design.

Both GMs are turned on as soon as this robot is powered on (see original code in Fig. 6.2), but the Trunk Extension is designed such that it would normally lift the Medium Tire (Port 1) off the ground surface so that its rotation does not interfere with the forward movement (Port 2) of the robot. However, if this Trunk Extension slips down an edge (of a table for example), the Medium Tire would then contact with the ground surface and make the robot turn left away from the ledge. This is a purely mechanical way to afford an "automatic" control system without much software costs. This technique is used in many motorized toys that the readers may have experience with already.

```
ENDLESS LOOP
{
    // Adjust value below to control the speed and direction
    PORT[1]:Geared Motor = CW:850 (83.09%)
    PORT[2]:Geared Motor = CW:300 (29.33%)
}
```

Fig. 6.2 Original Elephant Robot Code.

A "Smart Elephant" robot, using an NIR Sensor, is illustrated in Fig. 6.3, whereas the Trunk Extension is replaced by the NIR Sensor using Port 3 and small tires (O-rings) are also added for improved traction on the pulley-wheels of Port 2.

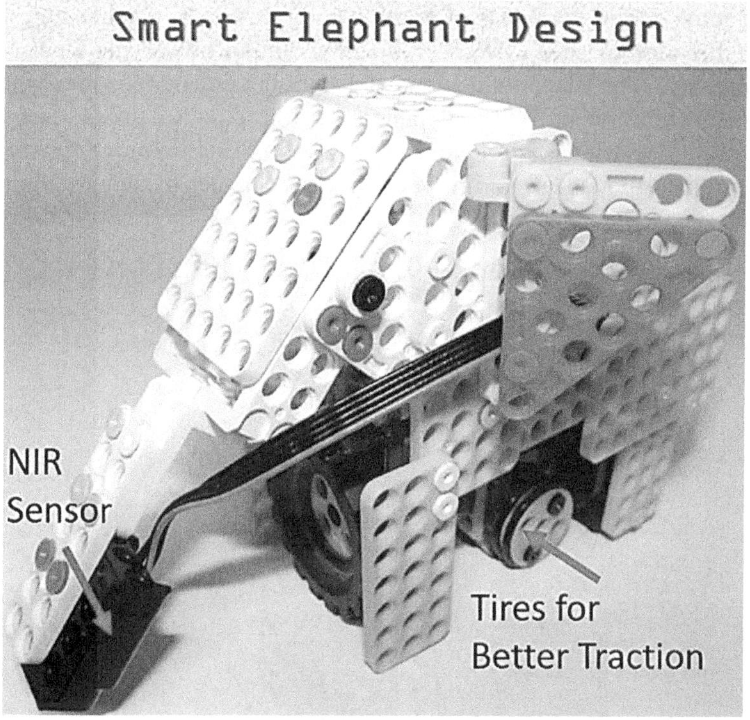

Fig. 6.3 New "Smart Elephant" Design.

The TASK program "RC_SmartElephant" uses a Remote-Control scheme (U-D only) combined with an automatic ledge avoidance algorithm. This code is based on "RC-MultiDirectionsSpeeds.tskx" (Section 3.7.2) with appropriate modifications to its directional movement Functions and the addition of the Ledge Avoidance Function.

Fig. 6.4 shows that it is only when the NIR Sensor reads in a value greater or equal to 25 (Line 29) that the CM-150 Controller would process the data input from the RC-100 into parameters ButtonUp, ButtonDown, Button1 and Button3 (Lines 32-35). Next, depending on which button is being pushed, the appropriate Function Calls are made (Lines 38, 42 or 46).

RC_SmartElephant.tskx (1)

```
IF ( PORT[3]:IR Sensor >= 25 )
{
    // Filtering out specific U-D-1-3 buttons
    ButtonUp = Data & U
    ButtonDown = Data & D
    Button1 = Data & 1
    Button3 = Data & 3
    IF ( ButtonUp == U )
    {
        CALL Forward
    }
    IF ( ButtonDown == D )
    {
        CALL Backward
    }
    IF ( Button1 == 1 )
    {
        Speed = 400
    }
```

```
ELSE
{
    CALL Ledge_Avoid
}
}
```

Fig. 6.4 Main Algorithm of "RC_SmartElephant.tskx".

The directional movements functions are modified appropriately as shown in Fig. 6.5, whereas Forward and Backward Functions are controlled by Port 2 while Left and Right Functions are controlled by Port 1.

RC_SmartElephant.tskx (2)

```
FUNCTION Forward
{
    PORT[2]:Geared Motor = CW:0 (0.00%) + Speed
}
FUNCTION Backward
{
    PORT[2]:Geared Motor = CCW:0 (0.00%) + Speed
}
FUNCTION Left
{
    PORT[1]:Geared Motor = CW:0 (0.00%) + Speed
}
FUNCTION Right
{
    PORT[1]:Geared Motor = CCW:0 (0.00%) + Speed
}
```

Fig. 6.5 Modified Directional Movement Functions in "RC_SmartElephant.tskx".

When the NIR Sensor reading is less than 25 (i.e. the ELSE on Line 57 in Fig. 6.4), the Function Ledge_Avoid (Line 59) is called (see Fig. 6.6). First the robot is stopped (Line 94). Next, the current Random Number is saved into Parameter RandomDirection. If RandomDirection has a value less or equal to 128, the robot goes backward and turn right (Lines 100-101), otherwise, the robot goes back backward and left. This is to simulate that 50% of the time, the robot will back away from the ledge and turn to the right, while for the other 50% of the time, the robot will back to the left.

Each "Ledge-Avoidance" move is controlled by a LOOP WHILE based on the NIR Sensor reading being less or equal to 85. This is to ensure that the robot is "safely" back inside the allowable ground surface before processing the user RC inputs once again. Of course, the reader needs to adjust the NIR limit values "25" and "85" to suit the reader's needs.

Video 6.1 shows the run-time performance of the program "RC_SmartElephant.tskx" on an "8-track" made of black electrical tape simulating a ledge off a table. This Video 6.1 shows that the Elephant was "hopping" while going forward but not when it is going backwards. This side effect is due to the Front Tire being set up perpendicular to the forward-backward direction, and when this tire rests on the ground surface, the friction between the tire and the ground surface impedes the robot's forward motion somewhat. This situation does not occur when the Elephant is going backwards, as this motion would lift the Front Tire slightly up and thus off the ground surface, so no hopping occurs. Video 6.1 also shows that at higher speed, the robot sometimes would run out of the black boundary, this just means that the track used should have a thicker width to better simulate an actual ledge off a table.

RC_SmartElephant.tskx (3)

```
92   FUNCTION Ledge_Avoid
93   {
94       CALL Stop
95       RandomDirection = 2 Random Number
96       IF ( RandomDirection <= 128 )
97       {
98           LOOP WHILE ( PORT[3]:IR Sensor <= 85 )
99           {
100              CALL Backward
101              CALL Right
102          }
103      }
104      ELSE
105      {
106          LOOP WHILE ( PORT[3]:IR Sensor <= 85 )
107          {
108              CALL Backward
109              CALL Left
110          }
111      }
```

Fig. 6.6 Function "Ledge_Avoid" in "RC_SmartElephant.tskx".

6.2 Flower & Butterfly with NIR Sensor

An NIR Sensor (Port 4) is added to the original robot design called "Flower & Firefly" to result in the "Flower & Butterfly" robot as shown in Fig. 6.7 (bottom right above the white LiPo battery). The author wanted to use a light source to simulate "daytime" (when this robot would operate) and "night-time" (when this robot would "sleep"). Although the TASK tool shows that an Illuminance Sensor is available in its "Set Device" menu (see Fig. 6.8), this sensor is not yet commercially available in the U.S.A., thus the NIR Sensor was used as a substitute sensor for this purpose. The TASK code changes are quite minimal as shown in Fig. 6.9 (essentially an IF-ELSE structure can do the job adequately – Lines 15 and 22). Video 6.2 shows the run-time performance of the program "FlowerButterfly_NIR.tskx".

Fig. 6.7 "Flower & Butterfly" with NIR Sensor used as Light Sensor (above battery).

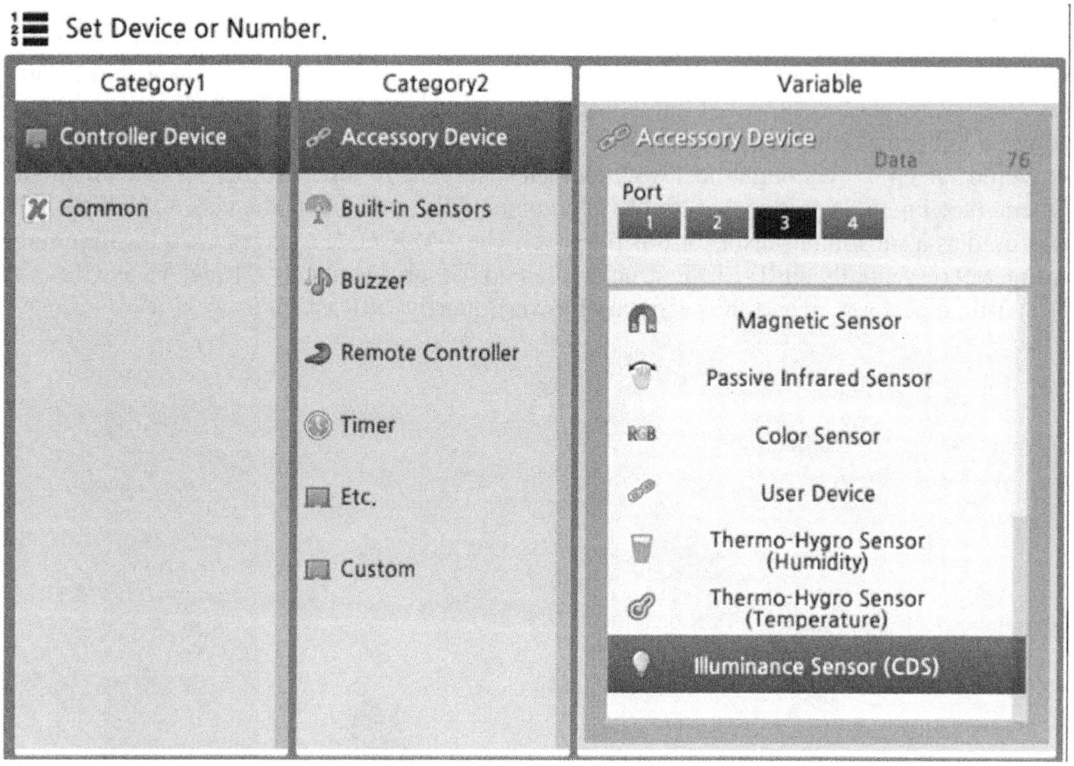

Fig. 6.8 Illuminance Sensor shown in TASK's Set Device Menu (but not yet available).

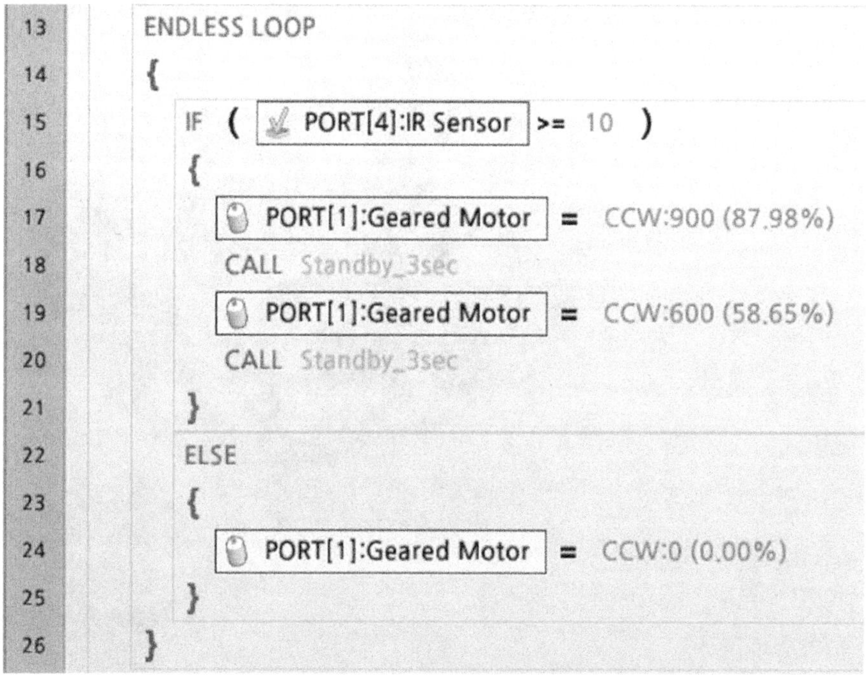

Fig. 6.9 Main Endless Loop in "FlowerButterfly_NIR.tskx".

6.3 Seal with Servo Head

In the original Seal Robot design, the head structure is mounted rigidly onto the robot's body. In this revised project, a Servo Motor (via Port 3) is mounted between the Seal body and its Head (see Fig. 6.10) to allow the robot to move its head from side to side. Please see Fig. 6.11 for details of the construction of the mount for the Servo Motor, and in previous steps the reader also needed to hook up the 5-pin cable to Port 3 before the assembly of the battery cradle.

Fig. 6.10 Seal Robot modified with Servo Head.

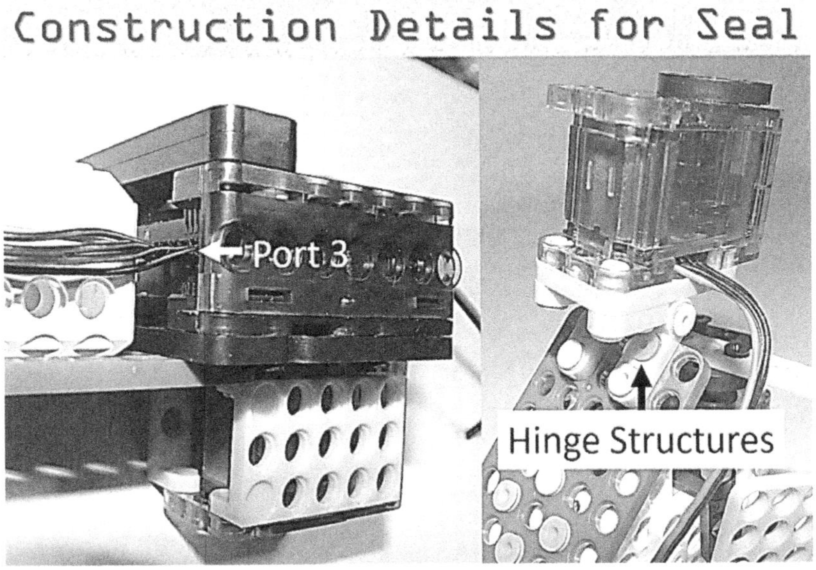

Fig. 6.11 Construction Details for Seal Robot modified with Servo Head.

The original TASK code was also modified slightly (see Fig. 6.12) so that the Head would move Left (Line 42) then Right (Line 47) before returning to the default Frontal Position (Line 52), for each time that the Seal was clapping its flippers. Video 6.3 demonstrates the run-time performance of the program "Seal_ServoHead.tskx".

```
30  ENDLESS LOOP
31  {
32      IF ( Result of Sound Counter > 0 )
33      {
34          Number_of_Repeated_Times = Result of Sound Counter
35          LOOP FOR ( n = 1 ~ Number_of_Repeated_Times )
36          {
37              // Plays melody each time the claps are detected.
38              Buzzer Timer = Melody Time
39              Buzzer Index = Melody No.13 (13)
40
41              // Claps and Moves Head.
42              PORT[3]:Servo Motor Position = 666
43              PORT[1]:Geared Motor = CW:400 (39.10%)
44              PORT[2]:Geared Motor = CW:400 (39.10%)
45              Timer = 0.512sec
46              WAIT WHILE ( Timer > 0.000sec )
47              PORT[3]:Servo Motor Position = 355
48              PORT[1]:Geared Motor = CCW:400 (39.10%)
49              PORT[2]:Geared Motor = CCW:400 (39.10%)
50              Timer = 0.512sec
51              WAIT WHILE ( Timer > 0.000sec )
52              PORT[3]:Servo Motor Position = 512
53              PORT[1]:Geared Motor = CCW:0 (0.00%)
54              PORT[2]:Geared Motor = CCW:0 (0.00%)
55              // Pauses and waited for the melody to stop.
56              Timer = 0.256sec
57              WAIT WHILE ( Timer > 0.000sec || Buzzer Timer > 0.0sec )
58          }
```

Fig. 6.12 Statements to move Seal's Head while it claps its flippers in "Seal_ServoHead.tskx".

6.4 Beetle with NIR Sensor

Back in Section 3.9.2, we had used an NIR Sensor attached to a servo motor which itself was mounted onto a wheel-based robot, assuring a stable platform, while the sensor was scanning for dowels (see Fig. 3.53). In this Section 6.4, we are going to explore the performance of a similar NIR dowel scanning system which has the NIR Sensor mounted onto a "walking" Beetle robot instead (see Fig. 6.13). Because of the 4-bar walking mechanisms used on its Left and Right sides, the Beetle rolls from side to side when it is in motion. This rolling motion in turn affects the NIR Sensor's view port and thus may affect the data stream used by the dowel search algorithm (see Fig. 3.55 and 3.59).

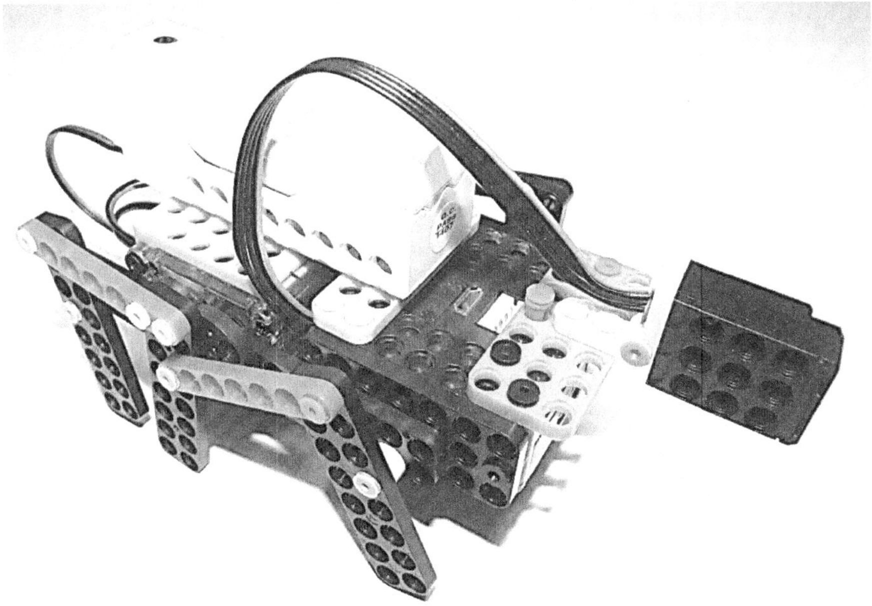

Fig. 6.13 Beetle Robot with NIR Dowel Scanning System.

For this Beetle project, the wanted operations for the robot are as follows:

1) The test wooden dowels need to be placed at random in front and to the right of the robot.

2) At start, the robot waits for the user's handclaps which correspond to the number of dowels that the robot is supposed to find.

3) Next, the robot swings right to find and approach the first dowel. When it gets close enough to the first dowel, it stops, plays a melody and waits for the user to clear the first dowel. Then it repeats the previous actions for the remaining dowels (one at a time).

The program "ScanningBeetle_NIR.tskx" was created to implement the above robot behaviors and it was derived from the "ScanningAvoider_CB.tskx" code. The algorithm used is described below:

1) For the first part of the Init Event, the robot plays Re# for 1 second (see Fig. 6.14 – lines 7-9), initializes various parameters (lines 10-17) and then waits for the user's handclaps (lines 18-19).

```
7   Buzzer Timer  =  1.0sec
8   Buzzer Index  =  Re# (6)
9   WAIT WHILE  ( Buzzer Timer  >  0.0sec  )
10  SpeedL  =  350
11  SpeedR  =  350
12  ProximityLimit  =  0
13  PreviousNIR  =  0
14  CurrentNIR  =  0
15  DowelDistance  =  0
16  ErrorCondition  =  0
17  FoundDowel  =  0
18  WAIT WHILE  ( Result of Sound Counter  ==  0  )
19  Claps  =  Result of Sound Counter
```

Fig. 6.14 Part 1 of the Init Event for program "ScanningBeetle_NIR.tskx".

2) Fig. 6.15 shows the second part of the Init Event whereas the controller reads in 20 consecutive data inputs from the NIR Sensor, i.e. "CB_NIR" (lines 20-24), computes the average and saves this value into "BackgroundNIR" (line 25). Next, "DowelThreshold" is set using "BackgroundNIR" (line 26), and the controller plays Melody 0 (lines 27-29). Please note that the CALLBACK Function is used to acquire the latest reading from the NIR Sensor which is attached to Port 4 (lines 149-152).

```
20  BackgroundNIR  =  0
21  LOOP FOR  ( I  =  1  ~  20  )
22  {
23      BackgroundNIR  =  BackgroundNIR  +  CB_NIR
24  }
25  BackgroundNIR  =  BackgroundNIR  /  20
26  DowelThreshold  =  BackgroundNIR  +  3
27  Buzzer Timer  =  Melody Time
28  Buzzer Index  =  Melody No.0 (0)
29  WAIT WHILE  ( Buzzer Timer  >  0.0sec  )
```

```
149  CALLBACK
150  {
151      CB_NIR  =  PORT[4]:IR Sensor
152  }
```

Fig. 6.15 Part 2 of the Init Event for program "ScanningBeetle_NIR.tskx".

3) Fig 6.16 shows the remainder of the Main Procedure whereas the controller executes a FOR LOOP for the same number of handclaps that the user provided in Step 1. Inside this FOR LOOP, the current dowel is located by Function "DowelScan" (see Fig. 6.17), next the robot gets closer to this dowel using Function "ApproachDowel" (see Fig. 6.19). If there is no error (i.e. ErrorCondition == 0), the FOR LOOP goes on to its next iteration (i.e. the next dowel). The "ErrorCondition" parameter is set inside the "ApproachDowel" Function and can have a value of "1" or "2" (see Fig. 6.20 for more details). If Line 34 of Fig. 6.16 evaluates to TRUE, the robot is stopped (line 36), then the program jumps to the LABEL named "Exit" (line 44), where it prints out "99999" followed by the actual value of "ErrorCondition".

```
84  FUNCTION DowelScan
85  {
86      DowelDistance = 0
87      CALL Right
88      // Finding Dowel
89      CALL FindDowel
90      DowelDistance = PreviousNIR
91      Print Screen with Line = DowelDistance
92  }
```

Fig. 6.16 Main Procedure for program "ScanningBeetle_NIR.tskx".

4) The details for Function "DowelScan" are shown in Fig. 6.17 whereas:

a) Parameter "DowelDistance" is initialized to zero (line 86) and the Beetle is commanded to turn right (line 87).

```
84  FUNCTION DowelScan
85  {
86      DowelDistance = 0
87      CALL Right
88      // Finding Dowel
89      CALL FindDowel
90      DowelDistance = PreviousNIR
91      Print Screen with Line = DowelDistance
92  }
```

Fig. 6.17 Function "DowelScan" of program "ScanningBeetle_NIR.tskx".

b) Function "FindDowel" is called on Line 89. This function uses the same dowel search algorithm explained in Section 3.9.2 (see Fig. 6.18 for details). Once returned from Function FindDowel, Parameter "PreviousNIR" contains the NIR Sensor reading corresponding to the latest "found" dowel and this value is saved into Parameter "DowelDistance" (line 90) and printed on the Output Screen by Line 91. The details for Function "FindDowel" are shown in Fig. 6.18 whereas:

The user is reminded that the robot was already turning right (line 87 of Fig. 6.17), by the time that Function "FindDowel" is called. Furthermore, at start, all the dowels were positioned on the right of the Beetle robot, i.e. the robot was staring into "empty space". Thus, the WAIT WHILE Loop at Line 96 is used to make the Controller "pass the time" until the edge of a dowel starts to appear within the viewport of the NIR Sensor (i.e. CB_NIR > DowelThreshold). At that time, the "actual" Dowel Search algorithm starts by assigning Parameters "PreviousNIR" and "CurrentNIR" with "logical" values. For "PreviousNIR", "BackgroundNIR" is the logical one to use (Line 97), because at this time, we know that the NIR Sensor starts to "see" the edge of the dowel so its values would be for sure greater than "BackgroundNIR". For "CurrentNIR", it is of course logical to assign the latest reading from the NIR Sensor to it, so Line 98 is used. "NIRDifference" is computed next (Line 99), and we know for sure that the WHILE LOOP on Line 100 would be entered next because of our previous choices for "PreviousNIR" and "CurrentNIR".

The loop defined between Lines 101 and 107 is used to swap the "CurrentNIR" value into "PreviousNIR" (Line 102) and to print it (Line 103). Next "CurrentNIR" is updated with the latest sensor reading "CB_NIR" and printed out (Lines 104-105). Finally, "NIRDifference" is updated and the loop's condition (NIRDifference > 0) is checked, and another iteration would be started if needed. Eventually, "NIRDifference" becomes negative when the dowel's centerline gets passed over, and the loop exits into Line 108 to print "PreviousNIR" (i.e. the actual "centerline" sensor reading). Finally, "FoundDowel" gets incremented by "1" and printed.

```
FUNCTION FindDowel
{
    WAIT WHILE ( CB_NIR <= DowelThreshold )
    PreviousNIR = BackgroundNIR
    CurrentNIR = CB_NIR
    NIRDifference = CurrentNIR - PreviousNIR
    LOOP WHILE ( NIRDifference >= 0 )
    {
        PreviousNIR = CurrentNIR
        [Print Screen] = PreviousNIR
        CurrentNIR = CB_NIR
        [Print Screen with Line] = CurrentNIR
        NIRDifference = CurrentNIR - PreviousNIR
    }
    [Print Screen] = PreviousNIR
    FoundDowel = FoundDowel + 1
    [Print Screen with Line] = FoundDowel
}
```

Fig. 6.18 Function "FindDowel" of program "ScanningBeetle_NIR.tskx".

5) The details for Function "ApproachDowel" are shown in Figs. 6.19 and 6.20. This function has error checking mechanisms using two IF structures (Lines 115 and 117 in Fig. 6.19), one on "DowelDistance" and the other on "CB_NIR", and normally these values need to be greater than zero to be "legitimate" values. Once these two conditions are satisfied, the robot can go forward towards the found dowel, but up to a certain distance only, by using the WHILE LOOP structure (Lines 119-122). Once CB_NIR exceeds "350" this loop is exited, and the robot stops (Line 123). Next, Melody 6 is played (Lines 124-126) and the robot is also made to wait for another 3 seconds (Line 127-128) so that the user has the time to remove the identified dowel from the test area (so that the robot can start searching for the next dowel as commanded by the number of handclaps made the user).

Fig. 6.20 shows how the possible error conditions (DowelDistance <=0, i.e. line 139) or (CB_NIR <=0, i.e. line 130) are conveyed to the user by playing Melody 9 and by setting the proper values "1" or "2" to Parameter "ErrorCondition".

```
FUNCTION ApproachDowel
{
    IF ( DowelDistance > 0 )
    {
        IF ( CB_NIR > 0 )
        {
            LOOP WHILE ( CB_NIR <= 350 )
            {
                CALL Forward
            }
            CALL Stop
            Buzzer Timer = Melody Time
            Buzzer Index = Melody No.6 (6)
            WAIT WHILE ( Buzzer Timer > 0.0sec )
            Timer = 2.944sec
            WAIT WHILE ( Timer > 0.0sec )
        }
```

Fig. 6.19 Part 1 of Function "ApproachDowel" of program "ScanningBeetle_NIR.tskx".

```
        ELSE
        {
            // Something went wrong
            Buzzer Timer = Melody Time
            Buzzer Index = Melody No.9 (9)
            WAIT WHILE ( Buzzer Timer > 0.0sec )
            ErrorCondition = 1
        }
    }
    ELSE
    {
        // Something went wrong
        Buzzer Timer = Melody Time
        Buzzer Index = Melody No.9 (9)
        WAIT WHILE ( Buzzer Timer > 0.0sec )
        ErrorCondition = 2
    }
```

Fig. 6.20 Part 2 of Function "ApproachDowel" of program "ScanningBeetle_NIR.tskx".

Video 6.4 demonstrates the run-time performance of the program "ScanningBeetle_NIR.tskx". This video shows that the "rolling" side effect of the "walking" feature of the Beetle can make it miss a dowel or result in a less accurate approach maneuver.

6.5 Racoon with PIR

The Passive IR Sensor (PIR-10) is a Motion Detector with a range of about 2 meters and a view angle of $\pm$ 45 degrees. In this application, it is mounted on the frame of a Racoon robot (see Fig. 6.21). Its output is "1" when motion is detected, and "0" when no motion is detected.

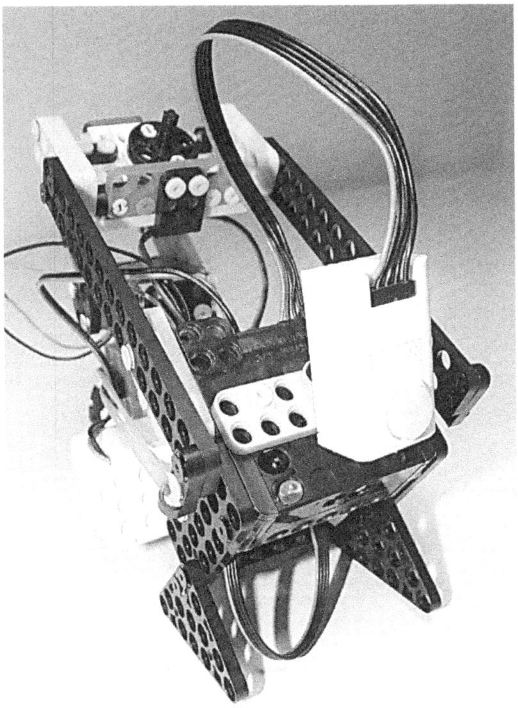

Fig. 6.21 Racoon Robot using PIR-10 Sensor as Motion Detector.

The application code "Racoon_PIR.tskx" is shown in Fig. 6.22 which consists of a Main Endless Loop whereas:

- Lines 17-20 is a WHILE LOOP designed as a visual aid to show the user that the Racoon's Motion Detection System is "live and armed" by printing the current value (= 0) of the CB_NIR Parameter (acquired via the CALLBACK Function).
- When a motion is detected, the loop is exited, and the Geared Motor on Port 1 is activated CW for 0.256 second to release the trigger mechanism of the catapult (lines 22-24).
- Next, the Port 1's Geared Motor is stopped and Melody 20 is played (lines 25-30).
- An extra wait time of 5 seconds is also used to allow the user to reset the catapult (lines 32-33).
- Finally, the robot waits for the user's handclap to restart the whole cycle once again.

Video 6.5 shows the run-time performance of the program "Racoon_PIR.tskx".

Racoon_PIR.tskx

```
ENDLESS LOOP
{
    // Robot stays still if there is no motion in its view port
    LOOP WHILE ( CB_PIR < 1 )
    {
        Print Screen with Line = CB_PIR
    }
    // Target found - shoot projectile
    PORT[1]:Geared Motor = CW:1023 (100.00%)
    Timer = 0.256sec
    WAIT WHILE ( Timer > 0.000sec )
    PORT[1]:Geared Motor = CW:0 (0.00%)
    Timer = 0.768sec
    WAIT WHILE ( Timer > 0.000sec )
    Buzzer Timer = Melody Time
    Buzzer Index = Melody No.20 (20)
    WAIT WHILE ( Buzzer Timer > 0.0sec )
    // Wait 5 seconds for user to reload catapult
    Timer = 5.120sec
    WAIT WHILE ( Timer > 0.000sec )
    // Wait for user clap
    WAIT WHILE ( Result of Sound Counter == 0 )
}

CALLBACK
{
    CB_PIR = PORT[4]:Passive Infrared Sensor
}
```

Fig. 6.22 Main Endless Loop in "Racoon_PIR.tskx".

6.6 Bull Fight using NIR and Color Sensors

The Bull/Buffalo Robot has its assembly instructions in the R+DESIGN tool under the DREAM I folder and in the Paper Manual that came with the DREAM II Level 2 kit. **The reader should also note that, for the Bull Robot, the R+DESIGN's instructions mount the drive motors in reverse from the Paper Manual's instructions** (at least currently in Spring 2018, the R+DESIGN information may be updated by ROBOTIS later). Thus the author created two versions of the Bull_Color codes: "Bull_Color_D.tskx" for R+DESIGN's users, and "Bull_Color_M.tskx" for Paper Manual's users.

The original Bull/Buffalo Robot uses the built-in Left and Right IR Sensors to detect and track an object, also it would charge forward when the detected object gets too close. This "revised" project additionally uses a Color Sensor (see Fig. 6.23) and illustrates a dual-range object detection scheme whereas:

- The NIR Sensors are used for the farther range (i.e. just to detect the presence of the object) and to make the robot move closer to the object.

• When the Color Sensor gets in range, then it is used to decide whether the object has a Red color or not. If a Red color is found, the Bull robot would back up and then charge ahead. For any other color found, the robot just plays a Melody.

The application code "Bull_Color_D.tskx" is shown in Fig. 6.24 which consists of a Main Endless Loop whereas:

• If the robot detects an object on its Left, it would turn left to track it (lines 20-21).

• As it gets closer to the object, the Left and Right IR Sensors would increase in value, and when both are larger than 100, the robot would next check the Color Sensor on Port 4 via the CALLBACK Function (i.e. get the latest Parameter CB_Color) (lines 23-25).

• If (CB_Color == Red) is TRUE, Function Charge is called (line 27), otherwise the robot would stop and Melody 21 is played (lines 31-34).

• Lines 38-55 (not shown) deal with the situation when an object is detected on the robot's Right in a similar manner.

Video 6.6 shows the run-time performance of the program "Bull_Color_D.tskx".

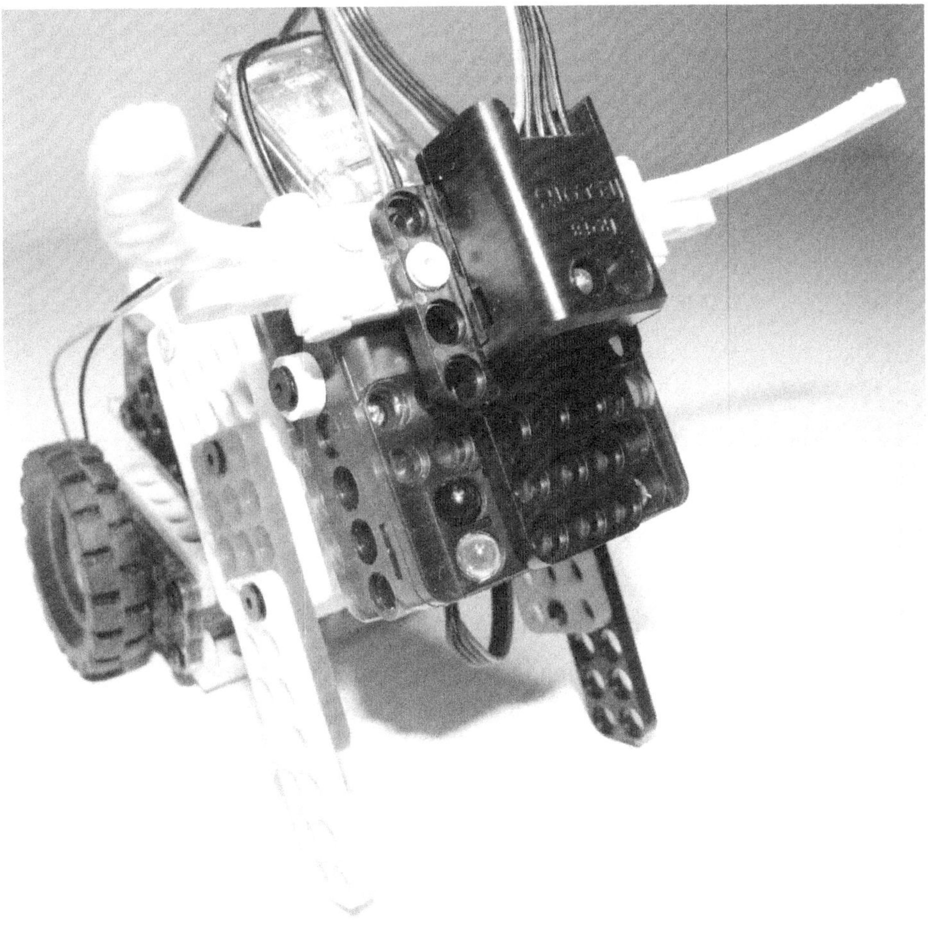

Fig. 6.23 Bull Robot with Color Sensor facing forward.

```
ENDLESS LOOP
{
    // Turns into the direction the object is detected. When both IR sensor:
    IF ( IR Left > 10 )
    {
        CALL Left
        IF ( IR Left >= 100 && IR Right >= 100 )
        {
            IF ( CB_Color == Red Color (3) )
            {
                CALL Charge
            }
            ELSE
            {
                CALL Stop
                Buzzer Timer = Melody Time
                Buzzer Index = Melody No.21 (21)
                WAIT WHILE ( Buzzer Timer > 0.0sec )
            }
        }
    }
    ELSE IF ( IR Right > 10 )
    {
        CALL Right
```

Fig. 6.24 Part 1 of Main Endless Loop in "Bull_Color_D.tskx".

6.7 Raging Bull - SCRATCH vs. TASK

In this Section, the goal is to contrast the logical solutions and the run-time performances of the same robot but using first a SCRATCH solution, and then a TASK solution.

We are using the previous Bull robot design in Section 6.6, but the Color Sensor had to be replaced by the NIR Sensor, as the Color Sensor is not supported by the R+SCRATCH/SCRATCH interface at present (see Fig. 6.25). However, we'll be using the audio-play feature of the SCRATCH interface to allow the Raging Bull to growl and to play "Ole!" exclamations from the "audience".

Please note that in the previous "Bull_Color.tskx" project (Section 6.6), the built-in Left and Right NIR Sensors were used as long-range sensors while the Color Sensor was used as short-range sensor. In this "Raging Bull" project, the Port 4 NIR Sensor is used as long-range sensor and the built-in Left and Right NIR Sensors are used as short-range sensors, and in the same manner as in the Follower projects of Sections 3.5 and 4.6.

Fig. 6.25 The "Raging Bull" Robot.

The "Init" Event for the SCRATCH Project "RagingBull_D1.sb2" is described in Fig. 6.26:

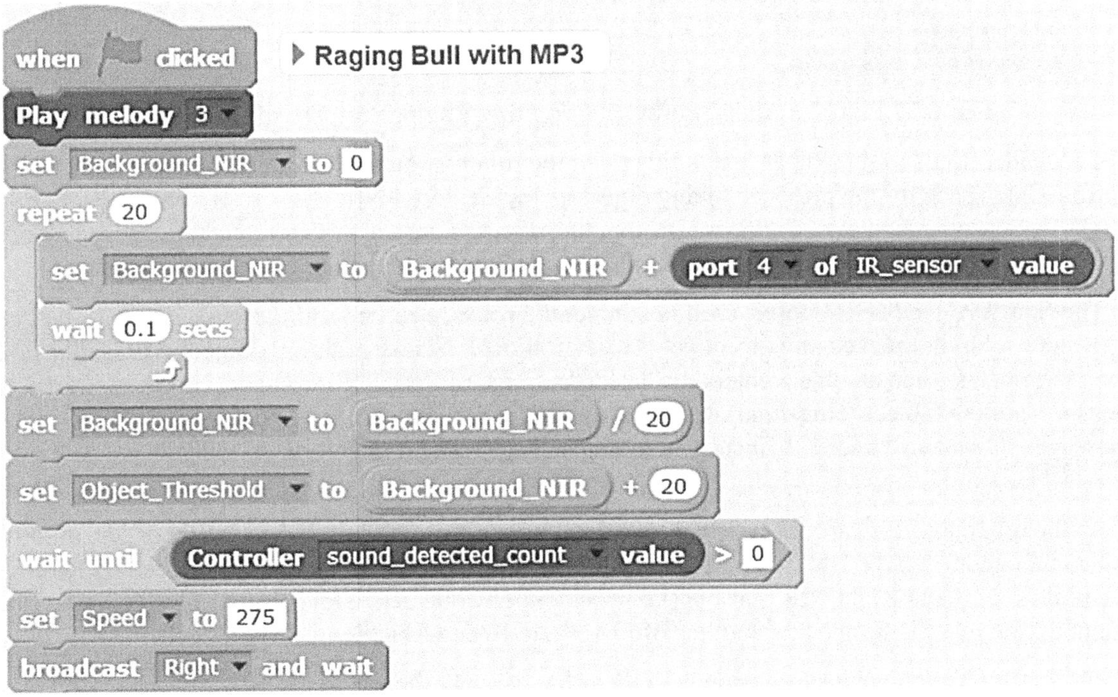

Fig. 6.26 "Init" Event for "RagingBull_D1.sb2" Program.

• First Melody 3 is played and Parameter "Background_NIR" is initialized to zero.

• Next a Repeat 20 loop is used to collect 20 consecutive values of the Port 4 NIR Sensor into an accumulating sum which is then divided by 20 to obtain its average value. This average value is finally saved into Parameter "Background_NIR". Parameter "Object_Threshold" is set as 20 units larger than "Background_NIR".

• Then the PC enters a "Wait Until" Loop, essentially "doing nothing" until the user makes a hand-clap. At that time, Parameter "Speed" is set to 275 and the Message "Right" is broadcasted to make the robot rotate right and in-place.

As usual, the main algorithm is implemented by a "Forever" Loop which starts as shown in Fig. 6.27:

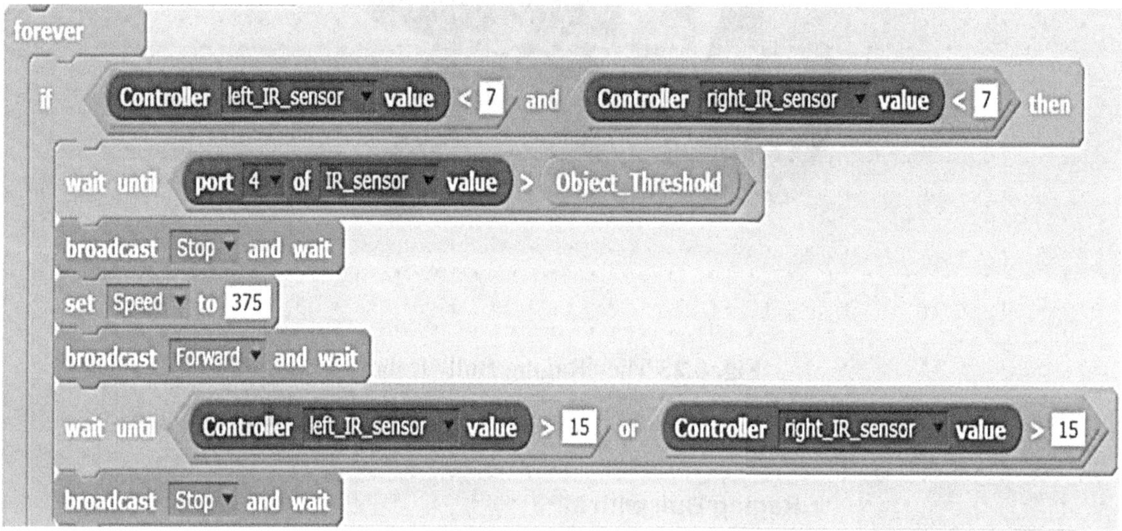

Fig. 6.27 Part 1 of Main "Forever" Loop for "RagingBull_D1.sb2" Program.

• The reader needs to remember that at this point the robot is rotating right and is constantly scanning its surrounding space for the presence of any object using the Port 4 NIR Sensor. If both built-in Left and Right IR Sensors read below 7, this means that the object, if any, is farther than the ranges allowed for the built-in NIR Sensors, thus the algorithm next concentrates on the use of Port 4 NIR Sensor.

• The first WAIT UNTIL Loop is used next to let the robot keep on turning right as long as the Port 4 NIR Sensor has not picked any object yet (i.e. (Port 4 of IR Sensor Value) < Object_Threshold). This loop terminates when an object enters the detection range of Port 4 NIR Sensor (i.e. when (Port 4 of IR Sensor Value) > Object_Threshold). Next the robot is stopped with a BROADCAST & WAIT "Stop" message. Parameter "Speed" is increased to 375 and the robot is sent forward with a BROADCAST & WAIT "Forward" message.

• As the robot goes forward towards the object previous detected by Port 4 NIR Sensor, the algorithm enters the second WAIT UNTIL Loop, essentially waiting for the object to enter the detection ranges of the built-in Left and Right NIR Sensors (i.e. when either value gets larger than 15). When this condition happens, the robot is stopped by another BROADCAST & WAIT "Stop" message.

From now on, the object is close enough to the robot, so only the built-in NIR Sensors are used to control the tracking of the object. This tracking algorithm is implemented by an IF-ELSE-IF structure described in Fig. 6.28:

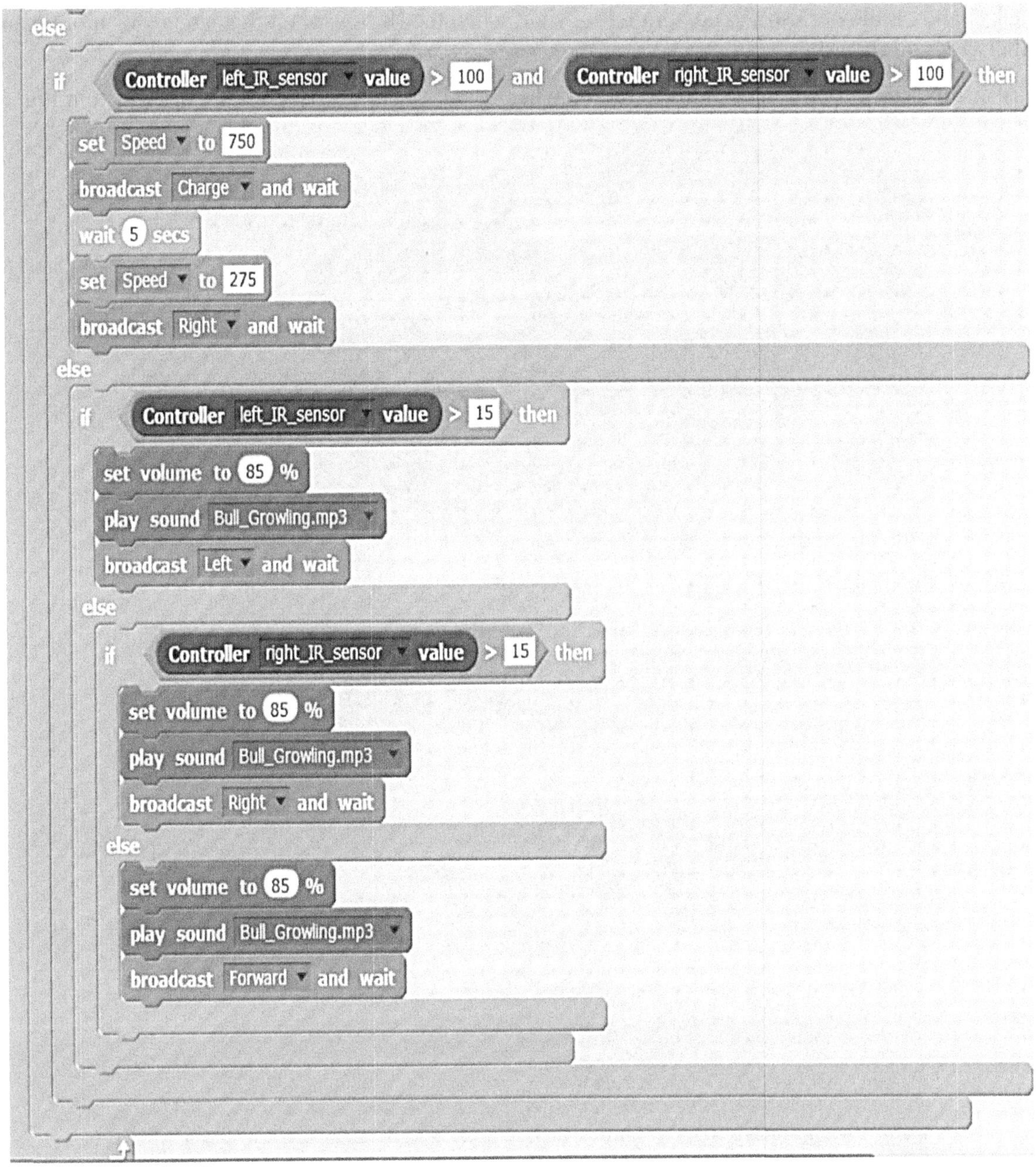

Fig. 6.28 Part 2 of Main "Forever" Loop for "RagingBull_D1.sb2" Program.

- If both built-in NIR Sensors read in values larger than 100, "Speed" increases once more to 750 and then the BROADCAST & WAIT "Charge" message is sent, triggering a series of actions described in Fig. 6.29. When these actions are done, a 5 second wait period is used to allow the user to reset the robot in the demonstration area. The "Speed" is reset to 275 and the robot is made to rotate right again looking for a new object (i.e. resume the FOREVER loop in Fig 6.27).
- The remaining IF-ELSE-IF structure is used to allow the robot to track the object if it happens to be moved around at random directions during this period (similar to the approach used in "Follower-

3.tskx" and "Follower3.sb2"), i.e. turn left or right when the IR Sensor on that side reads more than 15, otherwise go forward.

Lastly, the details for all the actions triggered when Event "Charge" happens are described in Fig. 6.29:

Fig. 6.29 Event "Charge" in "RagingBull_D1.sb2" Program.

- Volume is set to 100% and the Bull_Growling MP3 file is played while the robot goes backward for 0.25 second then stops for 0.25 second, and finally "charges" forward for 0.75 sec.
- Then the audience goes "wild" with an Olle!

The author also created a TASK version equivalent to this SCRATCH version where simple musical scales are used instead of the MP3 files:

- Fig. 6.30 describes the "Init" Event equivalent to Fig. 6.26.
- Fig. 6.31 describes Part 1 of the Main Endless Loop, equivalent to Fig. 6.27.
- Fig. 6.32 describes Part 2 of the Main Endless Loop, equivalent to Fig. 6.28.
- Fig. 6.33 describes Function "Charge", equivalent to Fig. 6.29.

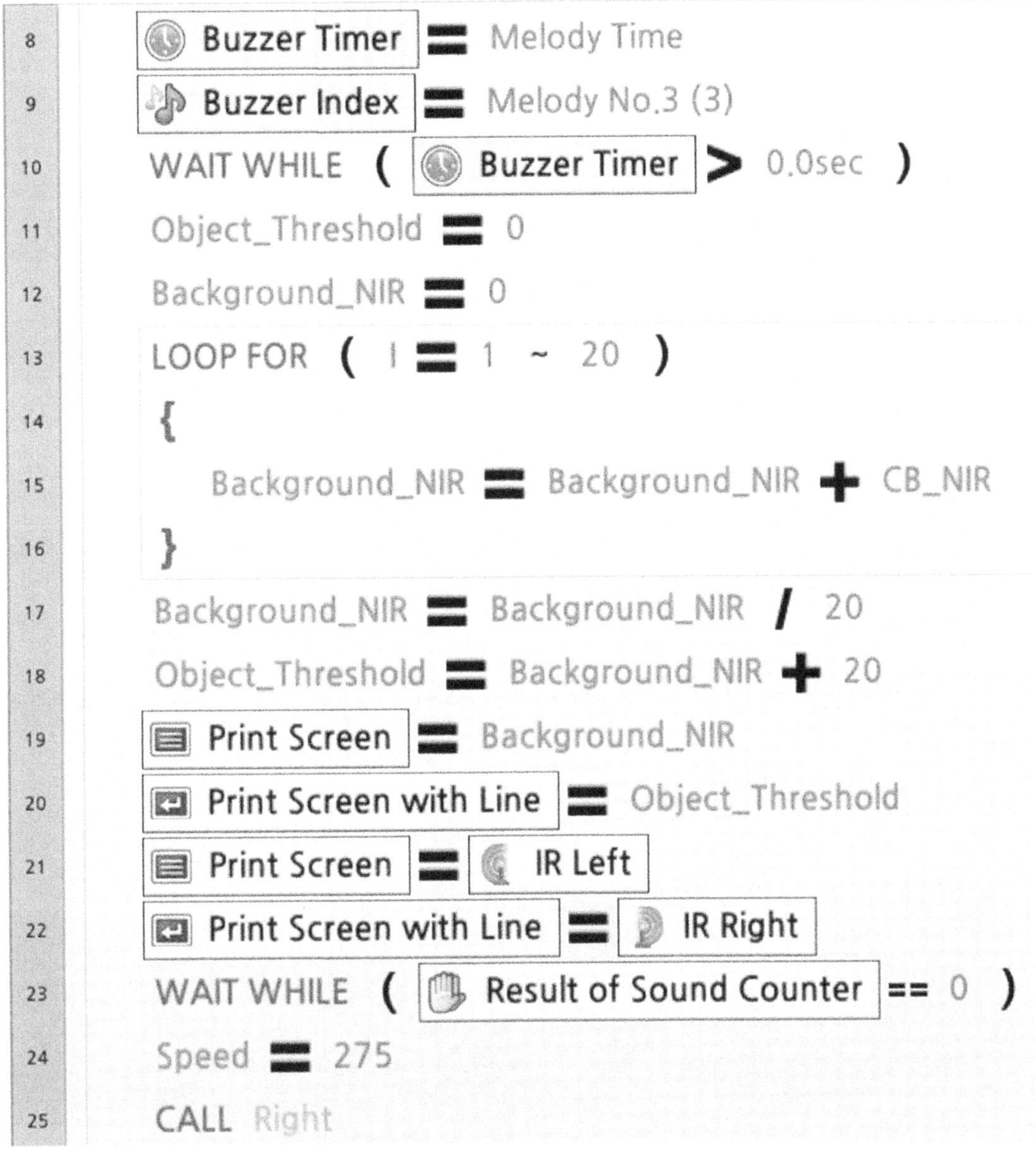

Fig. 6.30 "Init" Event in "RagingBull_D1.tskx" Program.

```
ENDLESS LOOP
{
    IF ( IR Left < 7 && IR Right < 7 )
    {
        WAIT WHILE ( CB_NIR <= Object_Threshold )
        CALL Stop
        Speed = 375
        CALL Forward
        WAIT WHILE ( IR Left < 15 && IR Right < 15 )
        CALL Stop
    }
```

Fig. 6.31 Part 1 of Endless Loop in "RagingBull_D1.tskx" Program.

```
ELSE
{
    IF ( IR Left >= 100 && IR Right >= 100 )
    {
        Speed = 750
        CALL Charge
        CALL Standby_5sec
        Speed = 275
        CALL Right
    }
    ELSE IF ( IR Left > 15 )
    {
        Buzzer Timer = 0.2sec
        Buzzer Index = La (12)
        CALL Left
    }
    ELSE IF ( IR Right > 15 )
    {
        Buzzer Timer = 0.2sec
        Buzzer Index = Sol (10)
        CALL Right
    }
    ELSE
    {
        Buzzer Timer = 0.2sec
        Buzzer Index = Re (17)
        CALL Forward
    }
}
```

Fig. 6.32 Part 2 of Endless Loop in "RagingBull_D1.tskx" Program.

```
FUNCTION Charge
{
    // Moves backward then charge
    CALL Backward
    CALL Standby_0_25sec
    CALL Stop
    CALL Standby_0_25sec
    CALL Forward
    CALL Standby_0_75sec
    CALL Stop
    Buzzer Timer = Melody Time
    Buzzer Index = Melody No.12 (12)
    WAIT WHILE ( Buzzer Timer > 0.0sec )
}
```

Fig. 6.33 Function "Charge" in "RagingBull_D1.tskx" Program.

Part 1 of Video 6.7 shows the run-time performance of the TASK program "RagingBull_D1.tskx" while Part 2 of Video 6.7 shows the run-time performance of the SCRATCH program "RagingBull_D1.sb2". It is very clear that the SCRATCH solution did not perform as well as the TASK solution due to the communication slowdown inherent in the SCRATCH2 >> R+SCRATCH >> CM-150 circuit, resulting in bad timing between sensor readings and associated maneuver executions. The reader can try tuning down the "Speed" of the Geared Motors to improve overall coordination.

Two other related SCRATCH projects are also included in the Example Codes for this book:

- In the "RagingBull_D2.sb2" solution, no MP3 files are played during the Left and Right tracking maneuvers. The author thought that this omission would improve the robot's performance, but it turned out to make the whole solution unworkable.
- The "RagingBull_D3.sb2" solution replaces all "BROADCAST & WAIT" structures with "BROADCAST" structures and this turns out to be a non-workable solution also.

6.8 MotoCycle

The MotoCycle Robot is newly introduced with the DREAM II School Set (see Fig. 6.34). It has a wide rear-drive wheel design (triple-tires) so that it can stand upright even when stopped (of course, not like a "real" motocycle). For this book's version, it has an additional external NIR Sensor (Port 4) mounted on the steering column to help it detect frontal obstacles when it is driven forward, while originally it only uses the built-in Center NIR Sensor to detect obstacles in its rear when it is driven backward. The steering column itself is controlled by a Servo Motor on Port 3.

Fig. 6.34 "MotoCycle" with wide rear drive wheel and NIR Sensor mounted on steering column.

The "MotoCycle.tskx" has an "Init" event described in Fig. 6.35:

```
9   Speed = 400
10  Buzzer Timer = Melody Time
11  Buzzer Index = Melody No.3 (3)
12  WAIT WHILE ( Buzzer Timer > 0.0sec )
13  PORT[3]:Servo Motor Drive Mode = TRUE (1)
14  PORT[3]:Servo Motor Speed = 1023
15  PORT[3]:Servo Motor Position = 512
```

Fig. 6.35 "Init" Event in "MotoCycle.tskx".

- Parameter Speed is set to 400 (line 9) and Melody 3 is played (lines 10-12).
- The Servo Motor on Port 3 gets set to its Position Control mode (line 13), its Speed (i.e. torque level) set the maximum 1023 value (line 14), and it is sent to its Front Facing position of 512 (line 15).

The main algorithm is implemented within an Endless Loop as usual, where Part 1 (see Fig. 6.36) sets up the RC procedure such that if the robot did not receive any user command, it would stop its Port 1 Geared Motor, i.e. the bot (line 24 in Fig. 6.36).

```
17  ENDLESS LOOP
18  {
19      WAIT WHILE ( Remocon Data Received == FALSE (0) )
20      DataIn = Remocon RXD
21      Print Screen with Line = DataIn
22      IF ( DataIn == -- )
23      {
24          PORT[1]:Geared Motor = CW:0 (0.00%) + 0
25      }
```

Fig. 6.36 Part 1 of Main Endless Loop in "MotoCycle.tskx".

Part 2 of the Main Endless Loop is described in Fig. 6.37:

- The user's RC input "DataIn" is parsed out into Parameters "FB_direction", "LR_direction" and "Mode" (lines 29-31).

- If the UP button is pushed (i.e. FB_direction == U), the MotoCycle rolls forward (line 44) if the Port 4 NIR Sensor reports a value smaller than 150 (i.e. no frontal obstacle). If there is a frontal obstacle, i.e. (Port 4 IR Sensor >= 150), then the MotoCycle is stopped (line 37), Melody 12 is played (lines 38-40) and a small Time Delay is also implemented (line 41).

```
ELSE
{
    // Filtering out specific Direction buttons or 1234 buttons
    FB_direction = DataIn & U+D
    LR_direction = DataIn & L+R
    Mode = DataIn & 1+2+3+4

    IF ( FB_direction == U )
    {
        IF ( PORT[4]:IR Sensor >= 150 )
        {
            PORT[1]:Geared Motor = CW:0 (0.00%) + 0
            Buzzer Timer = Melody Time
            Buzzer Index = Melody No.12 (12)
            WAIT WHILE ( Buzzer Timer > 0.0sec )
            CALL Delay_1
        }
        ELSE
            PORT[1]:Geared Motor = CW:0 (0.00%) + Speed
    }
```

Fig. 6.37 Part 2 of Main Endless Loop in "MotoCycle.tskx".

Part 3 of the Main Endless Loop is described in Fig. 6.38:

- If the DOWN button is pushed (i.e. FB_direction == D), the MotoCycle rolls backward (line 57) if the built-in Center NIR Sensor does not report any obstacle behind the robot. If there is a rear obstacle, i.e. (Center IR Sensor >= 100), then the MotoCycle is stopped (line 50), Melody 12 is played (lines 51-53) and a small Time Delay is also implemented (line 54).

```
46  ELSE IF ( FB_direction == D )
47  {
48      IF ( IR Center >= 100 )
49      {
50          PORT[1]:Geared Motor = CW:0 (0.00%) + 0
51          Buzzer Timer = Melody Time
52          Buzzer Index = Melody No.12 (12)
53          WAIT WHILE ( Buzzer Timer > 0.0sec )
54          CALL Delay_1
55      }
56      ELSE
57          PORT[1]:Geared Motor = CCW:0 (0.00%) + Speed
58  }
```

Fig. 6.38 Part 3 of Main Endless Loop in "MotoCycle.tskx".

Part 4 of the Main Endless Loop deals with Left and Right directional changes and is described in Fig. 6.39:

- If the LEFT button is pushed (i.e. LR_direction == L), the current Position of the Port 3 Servo Motor is increased by 50 (line 61), i.e. steering wheel rotating to the left.
- Else If the RIGHT button is pushed (i.e. LR_direction == R), the current Position of the Port 3 Servo Motor is decreased by 50 (line 63), making the steering column rotate to the right.

```
60  IF ( LR_direction == L )
61      PORT[3]:Servo Motor Position = PORT[3]:Servo Motor Position + 50
62  ELSE IF ( LR_direction == R )
63      PORT[3]:Servo Motor Position = PORT[3]:Servo Motor Position - 50
```

Fig. 6.39 Part 4 of Main Endless Loop in "MotoCycle.tskx".

Please note that as long the user pushes down on either "L" or "R" button, the steering wheel would keep on turning in the user-wanted direction, thus the steering wheel would only stop and stay in that position when the user releases that button.

The last part, Part 5, of the Main Endless Loop deals with various values for Parameter Mode (1, 2, 3 or 4). and it is described in Fig. 6.40:

```
65  IF ( Mode == 1 )
66  {
67      Buzzer Timer = Melody Time
68      Buzzer Index = Melody No.2 (2)
69      WAIT WHILE ( Buzzer Timer > 0.0sec )
70  }
71  ELSE IF ( Mode == 2 )
72      CALL IncreaseSpeed
73  ELSE IF ( Mode == 3 )
74  {
75      Buzzer Timer = Melody Time
76      Buzzer Index = Melody No.12 (12)
77      WAIT WHILE ( Buzzer Timer > 0.0sec )
78      CALL Delay_1
79  }
80  ELSE IF ( Mode == 4 )
81      CALL DecreaseSpeed
```

Fig. 6.40 Part 5 of Main Endless Loop in "MotoCycle.tskx".

- If (Mode == 1 or 3), the CM-150 plays Melody 2 at lines 67-69 or Melody 12 at lines 75-77, respectively.
- If (Mode == 2), Function IncreaseSpeed is called and if (Mode == 4), Function DecreaseSpeed is called.

Fig. 6.41 describes Functions "IncreaseSpeed" and "DecreaseSpeed" with the following features:

- The Upper Limit (i.e. 1000) and Lower Limit (i.e. 400) for Parameter "Speed" are controlled by its respective IF statement at line 94 or 105.
- Each Speed increase or decrease is by 200 units, and this change occurs only upon "release" of the buttons "2" or "4" by a clever use of the WAIT WHILE constructs at line 97 or line 108. This means that

the user needs to "press" either button "2" or "4" repeatedly to obtain multiple speed changes to the robot. Please contrast this approach to the one used for "L" and "R" buttons (see Fig. 6.39).

- Furthermore, there is an audio feedback each time that Parameter Speed gets changed (lines 98-99 and 109-110). This means that when Speed is set at its limits, 1000 or 400, the IF constructs at lines 94 and 105 are not executed, therefore no actual Speed change will happen no matter how many times the user is pressing the buttons "2" or "4", and the user will know about this situation because there won't be any audio feedback!

```
92  FUNCTION IncreaseSpeed
93  {
94      IF ( Speed < 1000 )
95      {
96          Speed = Speed + 200
97          WAIT WHILE ( Remocon RXD == 2 )
98          Buzzer Timer = 0.2sec
99          Buzzer Index = Sol (10)
100     }
101 }
102
103 FUNCTION DecreaseSpeed
104 {
105     IF ( Speed > 400 )
106     {
107         Speed = Speed - 200
108         WAIT WHILE ( Remocon RXD == 4 )
109         Buzzer Timer = 0.2sec
110         Buzzer Index = Sol (10)
111     }
112 }
```

Fig. 6.41 Functions "IncreaseSpeed" and "DecreaseSpeed" in "MotoCycle.tskx".

Video 6.8 demonstrates the various features of the MotoCycle Robot.

6.9 Four-Wheel Drive Dump Truck

This Dump Truck Robot is also a new example with the DREAM II system (see Fig. 6.42), in this book's version, it is enhanced with a rear-looking NIR Sensor which is used to trigger an Auto Dump procedure. It also features a 4 Wheel Drive concept using sprockets (see more details in Fig. 6.43). The actual Geared Motor's shaft is connected to the larger central sprocket which is engaged to 2 smaller sprockets which are connected to the actual wheels. Due this gear ratio transfer, the wheels rotate at a higher speed that the Geared Motors that drive them, making the Dump Truck the fastest robot in the example robots that can be built with the DREAM II School Set. This Dump Truck also features a carrier platform that can be raised or lowered using a Servo Motor on Port 3.

Fig. 6.42 4-Wheel Drive Dump Truck with rear NIR Sensor.

Fig. 6.43 Details of the 4-Wheel Drive design using sprockets.

The demonstration code for this Dump Truck is named "4W_Drive_DumpTruck.tskx" of which the "Init" Event is described in Fig. 6.44:

```
10 PORT[3]:Servo Motor Drive Mode = TRUE (1)
11 PORT[3]:Servo Motor Speed = CCW:1000 (97.75%)
12 PORT[3]:Servo Motor Position = 512
13 Speed = 1023
14 ServoPosition = 512
15 Buzzer Timer = Melody Time
16 Buzzer Index = Melody No.3 (3)
17 WAIT WHILE ( Buzzer Timer > 0.0sec )
```

Fig. 6.44 "Init" Event in "4W_Drive_DumpTruck.tskx".

- Proper operating parameters for the Port 3 Servo Motor are set by lines 10-12.
- Parameters Speed and ServoPosition are initialized to their default values (lines 13-14) and Melody 3 is played.

As usual, the overall algorithm is controlled by an Endless Loop of which Part 1 is illustrated in Fig. 6.45:

```
ENDLESS LOOP
{
  Timer = 0.768sec
  LOOP WHILE ( Remocon Data Received == FALSE (0) )
  {
    IF ( Timer == 0 )
    {
      PORT[1]:Geared Motor = CCW:0 (0.00%) + 0
      PORT[2]:Geared Motor = CCW:0 (0.00%) + 0
    }
  }
```

Fig. 6.45 Part 1 of Endless Loop in "4W_Drive_DumpTruck.tskx".

• Part 1 is designed to monitor the Remocon Data Received Flag (line 22) and when the robot has not received any new RC command within 0.768 sec (line 21), it will turn off power to both Port1 and Port 2 Geared Motors (line 26-27).

Part 2 of the Endless Loop is designed to handle the cases when the "UP" or "DOWN" buttons are pressed by the user (see Fig. 6.46):

• When the UP button is pressed (line 30), the two Geared Motors are energized to go forward (lines 32-33).

• When the DOWN button is pressed (line 35), the Port 4 NIR Sensor is checked next. If its reading is less than 200 (i.e. no low rear obstacle), the two Geared Motors are set to go backwards. However, if an obstacle is detected, it is assumed to be a Dump Bin, and the robot would stop (lines 39-40) and the Function AutoDump is called.

```
30  IF ( Remocon RXD == U )
31  {
32      PORT[1]:Geared Motor = CW:0 (0.00%) + Speed
33      PORT[2]:Geared Motor = CCW:0 (0.00%) + Speed
34  }
35  ELSE IF ( Remocon RXD == D )
36  {
37      IF ( PORT[4]:IR Sensor > 200 )
38      {
39          PORT[1]:Geared Motor = CCW:0 (0.00%) + 0
40          PORT[2]:Geared Motor = CW:0 (0.00%) + 0
41          CALL Auto_Dump
42      }
43      ELSE
44      {
45          PORT[1]:Geared Motor = CCW:0 (0.00%) + Speed
46          PORT[2]:Geared Motor = CW:0 (0.00%) + Speed
47      }
48  }
```

Fig. 6.46 Part 2 of Endless Loop in "4W_Drive_DumpTruck.tskx".

Part 3 of the Endless Loop is designed to handle the cases when the "LEFT" or "RIGHT" buttons are pressed by the user or when all buttons had just been released (see Fig. 6.47):

• When the LEFT button is pressed (line 49), the two Geared Motors are energized to make the robot rotate left (lines 51-52), and conversely when the RIGHT button is pressed (line 54), the two Geared Motors are energized to make the robot rotate right (lines 56-57).

• And when all buttons are released (line 59), power is turned off on both Geared Motors.

```
49  ELSE IF ( Remocon RXD == L )
50  {
51      PORT[1]:Geared Motor = CCW:0 (0.00%) + Speed
52      PORT[2]:Geared Motor = CCW:0 (0.00%) + Speed
53  }
54  ELSE IF ( Remocon RXD == R )
55  {
56      PORT[1]:Geared Motor = CW:0 (0.00%) + Speed
57      PORT[2]:Geared Motor = CW:0 (0.00%) + Speed
58  }
59  ELSE IF ( Remocon RXD == -- )
60  {
61      PORT[1]:Geared Motor = CCW:0 (0.00%) + 0
62      PORT[2]:Geared Motor = CCW:0 (0.00%) + 0
63  }
```

Fig. 6.47 Part 3 of Endless Loop in "4W_Drive_DumpTruck.tskx".

Part 4 of the Endless Loop is designed to handle the cases when the "2" or "4" buttons are pressed by the user who wants to increase or decrease Parameter Speed (see Fig. 6.48):

• The approach used is similar to the one used for the MotoCycle (Section 6.7), i.e. to clamp the Speed setting to an Upper Limit of 1000 or to a Lower Limit of 700, and to provide audio feedback when an actual Speed change occurs. The current approach is more verbose as it also provides some Print Screen outputs to provide the user with the current value of Parameter Speed. The extra steps in lines 69-70 and 83-84 are not really needed.

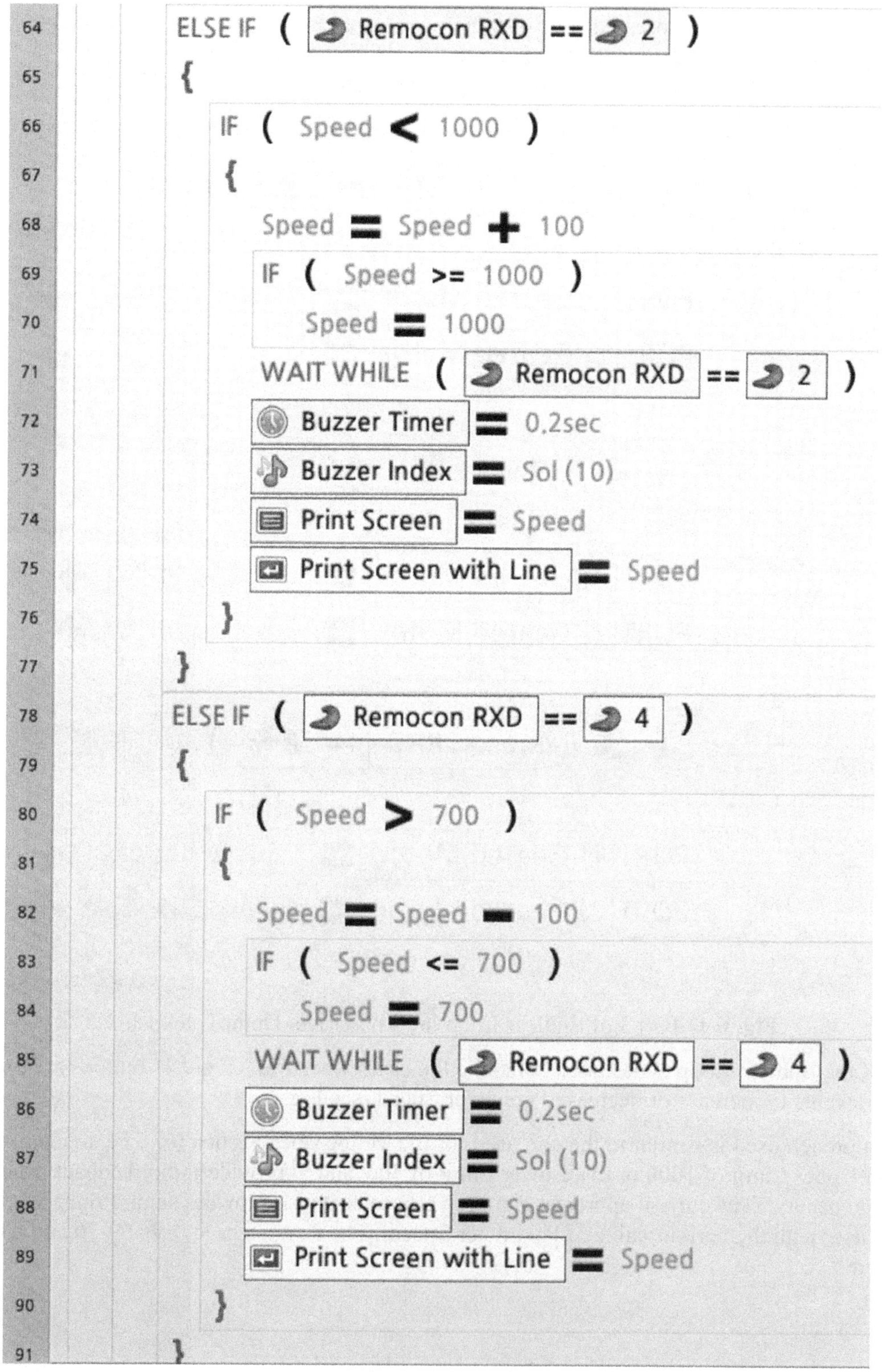

Fig. 6.48 Part 4 of Endless Loop in "4W_Drive_DumpTruck.tskx".

Part 5 of the Endless Loop is designed to handle the cases when the "1", "3", "5" or "6" buttons are pressed by the user (see Fig. 6.49):

- The user can press Button "1" (line 92) to raise the Dump Platform completely (i.e. Position 850 at line 93).
- The user can press Button "3" (line 94) to lower the Dump Platform completely (i.e. Position 512 at line 95).
- The user can press Button "5" (line 96) to raise the Dump Platform gradually by step of 50 (line 100) with a small Time Delay between each step.
- The user can press Button "6" (line 105) to lower the Dump Platform gradually by step of 50 (line 109) with a small Time Delay between each step.

```
92          ELSE IF ( Remocon RXD == 1 )
93            PORT[3]:Servo Motor Position = 850
94          ELSE IF ( Remocon RXD == 3 )
95            PORT[3]:Servo Motor Position = 512
96          ELSE IF ( Remocon RXD == 5 )
97          {
98            IF ( ServoPosition < 850 )
99            {
100             ServoPosition = ServoPosition + 50
101             PORT[3]:Servo Motor Position = ServoPosition
102             CALL Delay
103           }
104         }
105         ELSE IF ( Remocon RXD == 6 )
106         {
107           IF ( ServoPosition > 512 )
108           {
109             ServoPosition = ServoPosition - 50
110             PORT[3]:Servo Motor Position = ServoPosition
111             CALL Delay
112           }
113         }
```

Fig. 6.49 Part 5 of Endless Loop in "4W_Drive_DumpTruck.tskx".

Finally, the Function "AutoDump" is described in Fig. 6.50:

- First, the Dump Platform is fully raised (line 125).

- The sequence of Servo Position 750 (lines 127-128) followed by Servo Position 850 (lines 129-130) is designed to simulate a "shake" to clear carried materials that may still be stuck on the platform after the initial Dump.

- A second "shake" is performed by lines 131-134, followed by a final lowering of the Dump Platform to Position 512 (line 135), along with playing Melody 2.

```
117 FUNCTION Delay
118 {
119     Timer = 0.256sec
120     WAIT WHILE ( Timer > 0.000sec )
121 }
122
123 FUNCTION Auto_Dump
124 {
125     PORT[3]:Servo Motor Position = 850
126     CALL Delay
127     PORT[3]:Servo Motor Position = 750
128     CALL Delay
129     PORT[3]:Servo Motor Position = 850
130     CALL Delay
131     PORT[3]:Servo Motor Position = 750
132     CALL Delay
133     PORT[3]:Servo Motor Position = 850
134     CALL Delay
135     PORT[3]:Servo Motor Position = 512
136     Buzzer Timer = Melody Time
137     Buzzer Index = Melody No.2 (2)
138     WAIT WHILE ( Buzzer Timer > 0.0sec )
139 }
```

Fig. 6.50 Function "AutoDump" in "4W_Drive_DumpTruck.tskx".

Video 6.9 shows the run-time performance of this TASK program "4W_Drive_DumpTruck.tskx",

In parting, the author hopes that this book had been helpful in initiating the reader's Robotics ourney, may it remain joyful and rewarding for the reader in the future!

Made in the USA
Coppell, TX
05 November 2026

11307218R00111